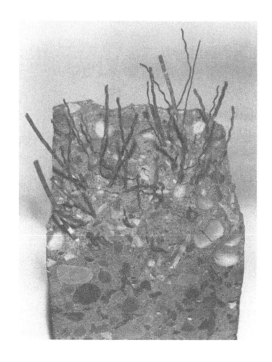

Fiber-Reinforced Cements and Concretes

Advances in Concrete Technology

A series edited by V. M. Malhotra
Scientist Emeritus
International Centre for Sustainable
Development of Cement and Concrete [ICON]
Ottawa, Ontario, Canada

This series will consist of approximately ten short, sharply focused tracts, each one covering one of the many aspects of concrete technology: materials, construction, and testing. The goal of this series is to provide a convenient, practical, and current source on concrete technology for practicing civil and structural engineers, concrete technologists, manufacturers, suppliers, and contractors involved in construction and maintenance of concrete structures.

Fiber-Reinforced Cements and Concretes

Colin D. Johnston

The University of Calgary
Calgary, Alberta, Canada

CRC Press
Taylor & Francis Group
Boca Raton London New York

CRC Press is an imprint of the
Taylor & Francis Group, an **informa** business
A TAYLOR & FRANCIS BOOK

CRC Press
Taylor & Francis Group
6000 Broken Sound Parkway NW, Suite 300
Boca Raton, FL 33487-2742

First issued in paperback 2019

ISBN-13: 978-90-5699-694-9 (hbk)
ISBN-13: 978-0-367-86451-4 (pbk)

Frontispiece: Fig. 1.1   Glass fiber-reinforced cement.
Fig. 1.2   Steel fiber-reinforced concrete.

British Library Cataloguing in Publication Data

Johnston, Colin D.
    Fiber-reinforced cements and concretes. – (Advances in
    concrete technology ; v. 3 – ISSN 1024-5308)
    1. Fiber-reinforced concrete 2. Fibrous composites
    3. Concrete construction
    I. Title
    624.1′8341

Visit the Taylor & Francis Web site at
http://www.taylorandfrancis.com

and the CRC Press Web site at
http://www.crcpress.com

CONTENTS

Contents   vii

PREFACE

Like other scientific and engineering disciplines, concrete
technology is subject to an increasing degree of specialization.
The incorporation of fibers into cementitious matrices to form
composites is an example of specialization within concrete
technology. It requires a knowledge of the concepts governing
interactions between the fibers and the cement paste, mortar or
concrete matrix that influence the manufacture and properties
of the end product. This volume is intended to summarize,
simplify and communicate the considerable body of research
and experience in fiber-reinforced cements and concretes. It is
ideal for engineers involved in design, construction or product
manufacture who are unfamiliar with this technology.

Three chapters deal with fundamental fiber-matrix interactions associated with the behaviour of randomly oriented discontinuous fibers in any cementitious matrix. In chapter 2
interactions involving the freshly mixed state, are discussed.
Interactions that determine composite behaviour in the hardened state, both in terms of early-age improvements and long-
term enhancement, follow in chapter 3. Reconciling these two
sets of often conflicting interactions is identified as the key to
producing optimal composite behaviour under load or other
conditions that induce tensile stress with the possibility of
cracking. Typical composite property improvements are
identified in chapter 4 where the scope is limited to composites with randomly oriented discontinuous fibers. It does not
cover the use of continuous unidirectional or two-dimensional
mesh and fabric forms of reinforcement.

Chapter 5 contains information on current practices with
respect to materials selection, mixture proportioning and manufacture for two distinctive categories of composite. The first
comprises thin-section, fiber-reinforced cements with relatively high fiber contents and significant two-dimensional fiber

alignment in the plane of the section. The fibers are often relatively fragile and are incorporated into matrices that do not contain coarse aggregate using manufacturing processes that minimize damage to the fibers. The second category comprises thick-section, fiber-reinforced concretes with lower fiber contents in matrices containing coarse aggregate. Fiber alignment is generally negligible except when the section thickness is less than about twice the fiber length. Manufacture is usually based on conventional mechanical mixing or shotcreting using relatively robust fibers capable of withstanding these processes without damage.

An overview of applications and potential applications for which the various composites may be appropriate is presented in the last three chapters. Significant property improvements achievable in practice and how they should be evaluated and specified are central to this discussion which draws upon the knowledge and experience gained from selected completed projects and the specifications on which they were based. Chapter 6 deals mainly with thin-section, fiber-reinforced cements that use alkali-resistant glass, carbon or natural and manufacture cellulose fibers and summarizes the more limited information on composites made with polyvinyl alcohol, acrylic, or polypropylene fibers. Chapter 7 deals with fiber-reinforced concretes, mainly in the context of steel or polypropylene fibers used in pavements, overlays and shotcrete. The more limited information on nylon or polyester fibers used to reduce plastic shrinkage cracking in slab-on-grade applications is also discussed briefly. Chapter 8 summarizes the key factors that need recognition to ensure success when using fibers in cement or concrete. The last three chapters provide owners, designers and users with the background needed to make informed decisions about the suitability of these composites for various applications.

ACKNOWLEDGMENTS

I am grateful to representatives of fiber manufacturers, equipment manufacturers, practising engineers with experience in specification development and construction practices, and academic colleagues involved in research. All have contributed in one way or another by reviewing and commenting on portions of the book relevant to their area of expertise, providing relevant technical papers, or supplying photographs of typical construction practices and products. In particular, my thanks to John Jones, Bud Molloy, Hiram Ball, Pete Tatnall, Dirk Nemegeer, Bob Zellers, Mel Galinat and David Gale for suggested refinements to the text to make it more accurately reflect current industry practice. Thanks also to Åke Skarendahl, Rusty Morgan, Bob Zellers, Pete Tatnall, John Jones, Geoff Osborne, Gary Mass, Graeme McGregor, Tony Naaman, David Lankard, Jens Jacob Niclassen and I.G. White for photographs. In addition, many of the above and others not already mentioned, including Rick Smith, Victor Li, Nemy Banthia, Simon Austin, David Hannant, Stefan Bernard and Ron Zollo, assisted me by providing reprints of relevant reports and technical papers.

Lastly, my thanks to René Kadach for patiently and meticulously typing the numerous drafts of the manuscript that led to the final text and to Bert Unterberger for capably preparing graphical and photographic illustrations.

CHAPTER 1

Introduction

The concept of reinforcing brittle building materials with various forms of fiber has been known since ancient times. Mud huts made using baked clay reinforced with straw and masonry mortar reinforced with animal hair are early examples of fiber-reinforced materials in construction. The first modern and sophisticated example of a fiber-reinforced construction material produced with a controlled and predictable level of quality was asbestos-cement, a material manufactured in thin-section forms such as flat or corrugated sheet and pipe by combining asbestos fibers with a slurry of cement and water, sometimes with finely divided silica, sand or other additives included.

Asbestos-cement manufactured by the Hatschek, Magnani or Mannville processes has been effectively and widely utilized in sheeting, roofing and cladding panels and in pipe since about 1900. In its final dewatered forms the fiber content by volume is 6–8% in sheeting, 8–10% in pipe, and 14–21% in fire-resistant boards, with the fibers typically less than 5 mm in length (Concrete Society 1973, Ryder 1975). From about 1970, its use has declined because of the hazard to human health now associated with breathing asbestos fibers, and considerable effort has been directed towards finding alternatives that are comparably effective for thin-section prefabricated applications in terms of engineering properties and cost. Like asbestos-cement, these newer fiber-cement composites with fibers such as glass (Fig. 1.1), carbon and aramid are characterized by relatively high fiber content, more than 2% by volume, and by a production process that blends the fibers into

1

FIGURE 1.1   Glass fiber-reinforced cement

a cement-based slurry without coarse aggregate thus avoiding
the possible damage that fragile fiber types suffer in a conventional concrete mixing process with coarse aggregate included.

A quite different category of fiber-reinforced cementitious
materials is the range of fiber-reinforced concretes made possible by including more robust discontinuous fibers as an ingredient of concrete in a conventional mixing process along with
other ingredients like aggregate and admixtures (Fig. 1.2). The
fiber content in these composites is much lower than in fiber-
cement composites, typically no more than 1.5% by volume
and sometimes as little as 0.1% by volume, and the fiber
length is longer, 15–65 mm. Apparently this idea originated
with a French patent in 1918 based on uniformly mixing small
longitudinal bodies (fibers) of iron, wood or other materials
into concrete (Naaman, 1985). The patent also suggested that
fiber elements must be rough, or be roughened, and that the
ends should be bent, features that are used today to improve

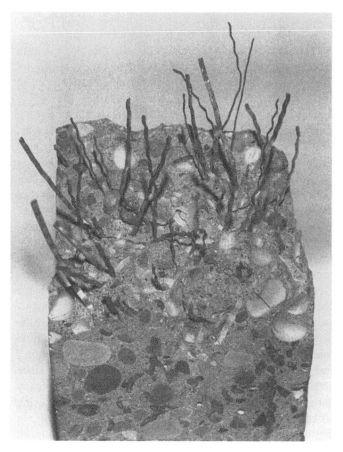

FIGURE 1.2   Steel fiber-reinforced concrete

the pullout resistance of fibers from concrete. Other patents
followed (Naaman, 1985), although the concept did not really
take hold until the 1960's when smooth straight steel fibers
produced by cutting wire or sheet metal became more widely
available commercially (U.S. Patent Office, 1969, 1970,
1972). In addition to steel, other types of fiber that have since
emerged specifically for use with conventionally mixed

concrete containing coarse aggregate include polypropylene, polyethylene and various types of polyester.

The reason for using fibers in both categories of composite is to enhance the properties of an inherently weak, brittle and crack-prone cementitious matrix. Depending on fiber type and fiber content, this enhancement may include in varying degrees improvements in tensile or flexural strength, ductility, toughness or energy absorption capability, impact resistance, fatigue resistance, resistance to cracking, permeability, and durability. However, amid the myriad of benefits claimed in available literature on the subject, particularly some of the promotional literature produced by fiber manufacturers, the user should recognize that the amount of fibers present is a major factor influencing the extent and degree of property enhancement. In this regard it is the volume fraction of fibers per unit volume of composite that is fundamentally important when comparing the effects produced by different types of fiber, even though it is convenient for practical purposes to batch fibers by weight and identify the fiber content in terms of weight per unit volume of composite. For example, it is realistic to compare the property enhancements obtained using fiber contents of 9.1 kg/m^3 of polypropylene (density 910 kg/m^3) and 79 kg/m^3 of steel (density 7860 kg/m^3) in concrete since both correspond to 1% by volume, and it would be technically unrealistic to compare the enhancement obtained using 40 kg/m^3 of polypropylene with 40 kg/m^3 of steel. It follows that the extent and degree of property enhancement is much greater in a fiber-reinforced cement with a fiber content of 5% by volume than in a fiber-reinforced concrete with 1% fiber, and is minimal or negligible in most respects in a fiber-reinforced concrete with 0.1% fiber.

1.1. TERMINOLOGY

The wide range of fiber-reinforced cementitious materials within the categories described above has led to some inconsistency in terminology mainly because the matrix containing the fibers may be a concrete, a mortar, or a cement paste. In its first state-of-the-art report in 1973 the American Concrete Institute defined *fiber-reinforced concrete* as "concrete made of hydraulic cements containing fine, or fine and coarse aggregate, and discontinuous discrete fibers". The current version of the report (ACI Committee 544, 1996) defines it simply as "concrete made primarily with hydraulic cement, aggregates, and discrete reinforcing fibers". Both versions imply the presence of aggregate and appear to exclude the category of fiber-reinforced cements. In 1977 the International Union of Testing and Research Laboratories (RILEM Committee 19-FRC, 1977) based in Europe defined *fiber concrete* as "made from hydraulic cements with or without aggregates of various sizes and incorporating, in the main, discrete fiber reinforcements". This definition includes the category of fiber-reinforced cements as well as mortars or concretes with fibers. Neither the ACI nor RILEM definitions adequately distinguishes between matrices consisting solely of cement paste, cement paste with fine aggregate (mortar), and cement paste with fine and coarse aggregate (concrete). The inconsistency is perpetuated in the terminology used to describe commercial products. For example, the material known as *glass fiber-reinforced concrete* or GFRC in N. America (ACI Committee 544, 1996, PCI Committee, 1993) is called *glass fiber-reinforced cement* or GRC in Europe (Majumdar and Laws, 1991) although it contains fine aggregate and is therefore really a fiber-reinforced mortar.

The same inconsistency in terminology is found in many technical papers purporting to discuss fiber-reinforced concrete, that on reading reveal that coarse aggregate was not included. Instead, the materials discussed are in fact mortars or cement pastes (often with particulate additives like fly ash or silica fume) that have properties far superior to their concrete equivalents because the nature of fiber-matrix interactions is such that the maximum fiber content possible in pastes and mortars is higher than for concretes.

1.2. NATURE OF FIBER-MATRIX INTERACTIONS

A wide variety of composites is possible when metal, glass, polymeric or natural fibers are composited with a matrix of cement paste, mortar or concrete. In the next two chapters important fiber-matrix interactions are identified in principle with minimal emphasis on specific fiber-matrix combinations in order that the reader acquire a broad understanding of the basic concepts before proceeding to chapters dealing with specific fiber-reinforced cements, mortars or concretes.

Fibers affect composite properties in both the freshly mixed and hardened states, often in opposite senses. For example, increasing the fiber content naturally tends to improve the degree of enhancement of many properties in the hardened state, but also decreases mixture fluidity in the freshly mixed state, until at some maximum fiber content the manufacturing process is no longer capable of producing uniform fiber distribution in a mixture that can be properly consolidated. This means that the potential enhancement of properties in the hardened state cannot be fully achieved either because of nonuniform fiber distribution or incomplete consolidation, or both. Another example is the conflicting role of fiber aspect ratio which is the ratio of fiber length to diameter for straight circular fibers. For reasons explained in the following chapters long slender high-aspect-ratio fibers offer greater reinforcing effectiveness and greater potential property enhancement in the hardened state than the same amount (fiber content) of short thick low-aspect-ratio fibers. However, high-aspect-ratio fibers also reduce mixture fluidity more severely than the same amount of low-aspect-ratio fibers. Consequently, mixtures with high-aspect-ratio fibers are more difficult to process in the freshly mixed state than mixtures with the same amount of low-aspect-ratio fibers, but offer greater potential for property enhancement in the hardened state if processed properly.

The reality is that low fiber content and low aspect ratio have the least adverse effect on properties in the freshly mixed state but offer little property enhancement, while high fiber content and high aspect ratio pose greater production difficulty but maximize the potential degree of property enhancement in the hardened state. Therefore, it should be recognized that what is possible in terms of miscibility, placeability, and final property enhancement is often a compromise reached after consideration of the various fiber-matrix interactions that affect composite behaviour in the freshly mixed and hardened states.

CHAPTER 2

Fiber-Matrix Interactions in the Freshly Mixed State

Fibers represent an addition of long slender needlelike particles to a normal cement paste, mortar or concrete matrix. Some like steel are relatively rigid, while others, like glass or polypropylene, are quite flexible. They can be added in many forms, as single filaments or monofilaments (Fig. 2.1, upper), as parallel permanently bonded multifilament strands with as many as 200 monofilaments in each unit (glass, Fig. 2.1,

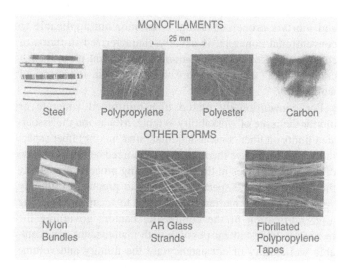

FIGURE 2.1 Fiber forms and types.

9

lower), as temporarily bonded bundles intended to separate into individual monofilaments during mixing (nylon, Fig. 2.1, lower), or as two-dimensional mesh-like fibrillated tapes intended to separate into branched monofilaments during mixing (fibrillated polypropylene, Fig. 2.1, lower). The characteristics of fiber-matrix combinations in the freshly mixed state depend on the type and form of the fibers, the nature and proportions of the matrix constituents, and the process used to incorporate the fibers into the matrix.

2.1. MIXTURE CONSISTENCY OR WORKABILITY

Adding any type of non-water-absorbent fiber to a cement paste, mortar, or concrete reduces the fluidity of the mixture because of the needlelike shape and high specific surface. Fibers which absorb water may cause further reduction in mixture fluidity. The behavior of cement pastes, pastes combining cement with fillers such as fly ash, slag, or silica fume, and mortars processed using techniques not applicable to conventional concrete, is usually characterized in terms of consistency. Concretes and mortars mixed and placed using conventional techniques are characterized in terms of workability.

For conventionally mixed fiber-reinforced concrete or mortar the issue of workability is more critical and the associated limits on fiber content and final (after mixing) fiber aspect ratio more restrictive than for fiber-reinforced cements because of major differences in the manufacturing process, principally the option in some special processes to prepare an initially very fluid paste or mortar and subject it to vacuum dewatering after fabrication of the composite. Factors governing the maximum fiber content possible with maintenance of acceptable workability or consistency are the fluidity and volume fraction of the paste phase, the maximum size of the coarse aggregate when present, and the fiber aspect ratio.

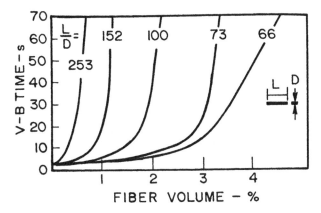

FIGURE 2.2   Effect of fiber aspect ratio on workability of steel
fiber-reinforced mortars (Hannant, 1978)

2.1.1. Effect of Fiber Aspect Ratio

For acceptable workability using placement by vibration, for
example V-B time of no more than 10s in Fig. 2.2 (Edgington,
Hannant and Williams, 1974), there is a well-defined limit on
fiber content for each aspect ratio beyond which workability
decreases sharply for the particular mortar tested. Such limits
will vary with the type of fiber and the proportions of the
mortar. What is acceptable workability obviously depends on
the nature of the application and the method of placement.

2.1.2. Effect of Paste Volume Fraction and Consistency

The greater the paste content, that is the volume fraction of
the fluid phase within which the fibers can move and rotate,
the greater the workability for any particular fiber content
(Fig. 2.3) (Peiffer and Soukatchoff, 1994), or in practice the
greater the amount of paste needed in the mortar to produce a
specified level of workability at the specified fiber content. Not
surprisingly, the intrinsic consistency of the paste phase is also
important, and in this case the determining factor is paste

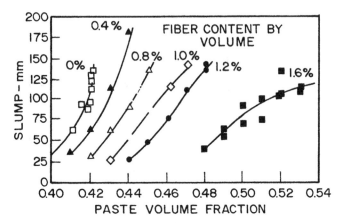

FIGURE 2.3   Effect of paste volume fraction, on workability of
steel fiber-reinforced mortars with 30 mm fibers
(Peiffer and Soukatchoff, 1994)

viscosity. This can be assessed by a flow test. The flow time or
consistency is governed by water-cement ratio, water-reducing
or superplasticizing admixture dosage, and the amount and
type of filler such as fly ash, slag or silica fume. For a specific
type and amount of fiber, the volume of paste needed to
produce a specified level of workability in fiber-reinforced
mortar decreases as the consistency or fluidity of the paste,
reflected by the flow time in Fig. 2.4, increases (Peiffer and
Soukatchoff, 1994). However, excessive fluidity, such as might
result from superplasticizer overdose, can cause segregation
and loss of mixture workability due to particle interference,
just as for mortar without fibers.

2.1.3. Effect of Aggregate Maximum Size

Pastes and mortars can accommodate relatively high fiber contents because of the relatively high volume fraction of the fluid
phase within which the fibers must be confined. In normally
proportioned concretes the volume fraction of the fluid phase

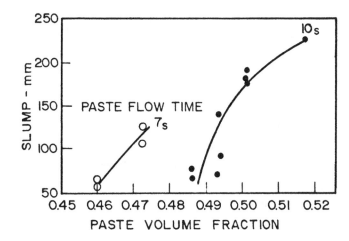

FIGURE 2.4 Effect of paste consistency (funnel flow time) and volume fraction on workability of steel fiber-reinforced mortars with 1% by volume of 30 mm fibers (Peiffer and Soukatchoff, 1994)

decreases with increases in the volume fraction and maximum size of the aggregate, so the volume of the space available for fibers decreases correspondingly. A 2-dimensional illustration (Fig. 2.5) of what is of course a 3-dimensional reality has been proposed (Hannant, 1978), presumably for straight rigid fibers. It shows why the fiber content corresponding to a specified level of workability, for example a V-B time of 10s, is much less for 20 mm aggregate concrete than for 10 mm aggregate concrete or mortar (Fig. 2.6) (Edgington, Hannant and Williams, 1974). Obviously, such limits depend on fiber type and aspect ratio. Whether flexible fibers bend around the aggregate particles sufficiently to significantly increase the maximum fiber content corresponding to any particular level of workability is unknown since rigid and flexible fibers that are geometrically identical have not apparently ever been compared in this regard.

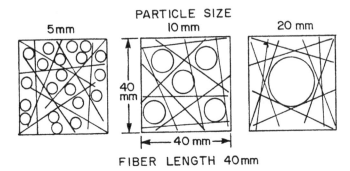

FIGURE 2.5   Schematic of particle size vs. fiber distribution for
              40 mm long fibers within a 40 mm square (Hannant,
              1978)

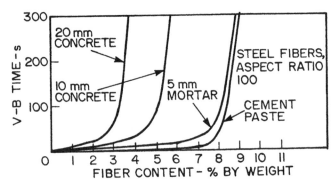

FIGURE 2.6   Effect of aggregate maximum size on workability for
              steel fibers of aspect ratio 100 (Hannant, 1978)

2.2. COMPATIBILITY WITH THE MANUFACTURING PROCESS

To be effective as reinforcement the fibers must not be
significantly damaged by the manufacturing process used to
incorporate them into the cementitious matrix. These
processes vary considerably in terms of the severity of damage
inflicted on the fibers by bending, impact and abrasion. Some

alter the fiber form in an acceptable or intended way by separating bundles or fibrillated forms into individual uniform or
branched monofilaments (Fig. 2.1, lower). Others can cause
unintended damage such as breakage that produces shortening
and loss of reinforcing effectiveness due to decreased aspect
ratio, or in the case of multifilament strands separation into
monofilaments that greatly increases specific surface with a
very adverse effect on workability or consistency and sometimes clumping of the monofilaments into balls in the mixture.

2.2.1. Spray Process for Cements, Pastes or Mortars

Least severe in terms of potential damage to the more fragile
types of fiber such as glass and carbon are processes that avoid
mechanical mixing entirely. In the spray process, chopped
fibers and a premixed cementitious paste or mortar slurry are
sprayed, usually from separate nozzles, onto a molding surface
where they mix as they impinge on the surface (Fig. 2.7), so
the possibility of damage to the fibers is minimal because they
never enter a mechanical mixer or come into contact with
coarse aggregate (Majumdar and Laws, 1991).

2.2.2. Premix Process for Cement Pastes or Mortars

The two-stage premix process involving preparation of the slurry
in a mixer followed by blending of the fibers into it after the
slurry has been mixed is another alternative for fragile fibers
which minimizes the time they are exposed to potential damage
by the mixer. However, experience has shown that tangling or
balling of the fibers with consequent nonuniform fiber distribution and difficulties with handling and placement can be a
problem that must be addressed by use of dispersing additives
and proper selection of the type of mixer. In the case of glass
fiber-reinforced composites produced at the U.K. Building
Research Establishment (Majumdar and Laws, 1991) a pan
mixer was found preferable to a drum mixer for the fiber addition stage, and a two-mixer process was eventually adopted with

FIGURE 2.7   Mechanical spray process for glass fiber-reinforced
cement pastes or mortars (Courtesy of U.K. Building
Research Establishment)

the slurry premixed in a high shear mixer before being transferred to the pan mixer. Various special mixers have been developed to facilitate mixing for fragile fibers, such as the high shear
mixer (Fig. 2.8) for use with glass fibers (Peter, 1994) and the
high energy, flexible-base Omni-mixer for use with carbon fibers
(Akihama, Suenaga and Banno, 1982) (Concrete Construction,
1986). Additives such as methyl cellulose and polyethylene
oxide may also be used to facilitate fiber dispersion.

2.2.3. Modified Hatschek Process for Cement Pastes or
Mortars

Developed originally for the manufacture of asbestos-cement
products, this process or modifications of it have been exam-

FIGURE 2.8   High shear (1400 rpm) mixer used to prepare mortar
prior to adding glass fibers in the premix process
(Courtesy of Power-Sprays Ltd.)

ined in the search for fibers to replace health-hazardous
asbestos in a line of products similar to the flat and corrugated
sheets and pipes historically produced in asbestos-cement. It
involves premixing of the fibers with a dilute cementitious
slurry, pick-up of a thin film of fiber-reinforced slurry by a
rotating cylinder of fine wire mesh, transfer of the film to an
endless conveyor belt of permeable felt, vacuum dewatering
of the film, and accumulation of successive films on a drum
combined with pressure compaction and further dewatering
until the required thickness is achieved (Fig. 2.9) (Ryder,

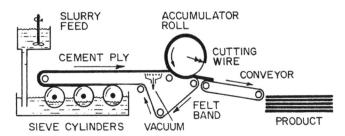

FIGURE 2.9   Hatschek process for manufacture of fiber-cement
             sheets (Gale, 1994)

1975) (Majumdar and Laws, 1991) (Gale, 1994). It is
best suited for lower density fibers that are light enough to
remain in suspension in relatively dilute slurry and flexible
enough to perform well during film forming, for example
polyvinyl acetate and wood pulp (Coutts, 1992). The capability of the fibers to successfully act as filters to facilitate
transfer of the smallest of the cement particles from the
slurry to the film, thus avoiding the accumulation of dead
cement (cement that has been in slurry so long that its ability
to hydrate has been lost) in the mixer/agitator, is also important since most fibers are less effective in this regard than
asbestos (Gale, 1994).

2.2.4. Conventional Mixing Process for Mortars or Concretes

More severe in terms of potential damage to the fibers are the
various conventional concrete mixing processes where the
rotation of the mixer, the shearing action of the blades, and
contact with coarse aggregate particles subjects fibers to considerable bending, impact and abrasion. Fragile fibers are
unsuitable for these processes because they sustain breakage
causing loss of aspect ratio and consequent loss of reinforcing
effectiveness. Multifilament strands like glass that are not
intended to separate tend to do so, and render the mixture

unworkable because of their greatly increased surface area. The performance of fibers in the form of fibrillated tapes like polypropylene that are intended to separate or defibrillate during mixing into individual monofilaments with secondary branches, or in the form of bundles like polyester that are also intended to separate into individual monofilaments, is influenced by the mixer type and mixing process. Insufficient mixing may fail to achieve the intended degree of separation. Overmixing may lead to filament breakage with consequent loss of aspect ratio, and possibly the loss of the secondary branching in fibrillated types needed for mechanical bond. Optimal mixing time is necessary to achieve satisfactory workability without damage to fibers and consequent loss of reinforcing effectiveness (Fig. 2.10) (Sakai et al., 1994).

For conventionally mixed fiber-reinforced concretes, and also wet-process shotcrete, workability and placeability are generally more critical than for the fiber-reinforced cement-based composites prepared by the spray or premix processes

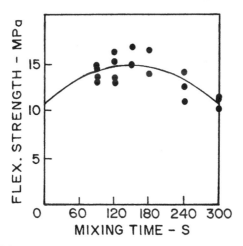

FIGURE 2.10   Relationship between composite strength and mixing time in a conventional mortar mixer for carbon fiber-reinforced cements (Sakai et al., 1994)

where dewatering after placement is common and allows excess water to be used during mixing. However, proper use of water-reducing and superplasticizing admixtures can allow the workability of fiber-reinforced concretes to be improved without employing excess water.

2.2.5. Shotcreting Processes for Mortars or Concretes

Like conventional mixing, shotcreting requires consideration of the potential for fiber damage. For the wet process, the considerations are the same as for conventional mixing. The dry process avoids damage due to mixing. In both the dry and wet processes, shooting against a hard, often rough, surface may inflict further damage to the more fragile fiber types, resulting in loss of aspect ratio. Therefore, only the more robust fibers like steel or polypropylene are suitable for either process. Regardless of fiber type, the primary concern in shotcreting is loss of in-place fiber content and consequently reduced reinforcing effect-iveness due to rebound. Rebound tends to be less severe in the wet process than in the dry process, so the wet process has tended to become more favoured by experienced users.

To improve fiber reinforcing effectiveness beyond what is possible in the normal dry or wet shotcreting processes, where fiber length must be kept relatively short to avoid problems with excessive loss of workability or fiber balling, a special process capable of accommodating longer fibers has been developed. This involves separate feeding of fiber material and matrix to the nozzle where the fibers are cut into lengths up to 200 mm and emitted concentrically within a cylindrically-shaped flow of matrix. The special spray gun for use with steel fibers incorporates two feeds of steel wire fiber into a flow of pumped concrete and compressed air (Skarendahl, 1992).

2.2.6. Slurry Infiltration Process For Pastes or Mortars

The slurry infiltration process was developed primarily to permit much higher fiber contents than are possible with any of the conventional mixing processes. It involves placement of the fibers on a forming surface in a manner that achieves acceptably uniform distribution followed by infiltration of the preplaced fibers with a slurry of paste or mortar sufficiently fluid to penetrate fully through the fiber network (Fig. 2.11) (Lankard, 1984). Obviously, a secondary advantage is that there is virtually no potential for damage to the fibers. However, the range of fiber contents possible is limited and is determined largely by the type and geometry of the fiber and the placement technique used to fill the mold or formwork. The process has been utilized primarily with steel fibers at fiber contents up to 20% by volume (Lankard, 1986).

FIGURE 2.11   Slurry infiltration process using steel fibers
(Courtesy of D.R. Lankard)

CHAPTER 3

Fiber-Matrix Interactions in the Hardened State

Fibers in hardened cement paste, mortar or concrete may have at least three important effects. First, they may tend to increase the stress at which the matrix starts to crack. This strengthening effect is most evident under modes of loading which induce tensile stress, for example direct tension, bending and shear. The degree of strengthening imparted by the fibers varies widely from insignificant (less than 10%) to substantial (several times the strength of the parent matrix in direct tension or bending) depending on the mode of loading and the type and amount of fibers.

Second, and again depending on the mode of loading and the type and amount of fibers, they may improve the strain capacity or ductility of the inherently brittle cementitious matrix, thus increasing its energy absorption capability or toughness characterized in general by the area under a stress-strain or load-deformation curve or some defined portion of it. Improvements in ductility and toughness are usually substantial even when improvements in strength are minimal. This toughening effect applies also to modes of loading other than those that induce tensile stress, in particular uniaxial compression where strengthening is often negligible.

A third important effect of fibers is their tendency to inhibit or modify crack development in terms of reducing crack width and average crack spacing. Again, the degree of improvement depends on fiber type and amount, together with the nature of the crack-inducing mechanism, for example, stress-controlled conditions produced by direct loading, or strain-controlled conditions caused by restrained shrinkage or thermal contraction associated with drying or temperature decrease.

With regard to strengthening, toughening and resistance to cracking, the reinforcing effectiveness of multiple randomly oriented fibers distributed uniformly throughout the matrix can be explained to some extent in terms of the intrinsic fiber properties and the factors which determine the shear bond at the interface between a single fiber and the matrix. These factors govern the resistance of the fiber to pullout from the matrix without which there would be no improvements in composite performance. The resistance to interfacial shear and fiber pullout may involve adhesion, friction and mechanical interlock. When the interfacial shear resistance is high enough the fiber breaks instead of pulling out of the matrix.

3.1. FIBER REINFORCING EFFECTIVENESS

The factors which determine the reinforcing effectiveness of fibers in a cementitious matrix are initially easiest to understand for the condition of uniform direct tensile stress in the composite. The concepts applicable to direct tension facilitate subsequent understanding of composite behaviour under other modes of loading that are more important in practical applications, such as bending and shear where the normal and principal stresses are in part tensile and are not necessarily uniform. The fiber-matrix interaction is also more easily understood if it is considered initially in terms of a single fiber aligned in the direction of composite tensile stress with stress transferred from matrix to fiber by interfacial shear (Fig. 3.1, left). Subsequently the effects of oblique fiber orientation and group action of multiple fibers can be incorporated into the analysis.

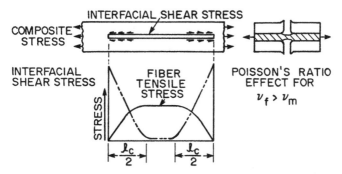

FIGURE 3.1   Single aligned fiber in composite under direct tension
(Broutman and Krock, 1967) producing interfacial
shear (left) and transverse stress due to Poisson's ratio
differential (right) (Kelly, 1975)

3.1.1. Intrinsic Fiber Properties

Any type of fiber effective for reinforcing relatively weak and
brittle cementitious matrices must have appropriate intrinsic
material properties as follows:

Tensile Strength – The fiber must be much stronger than the
matrix since the effective load-bearing area of a typical fiber
amount in the composite, 1–5% by volume, is much less than
the corresponding area for the matrix, 95–99%.

Ductility or Elongation – The fiber must be able to withstand
strains well in excess of the matrix cracking strain in order to
provide significant toughening.

Elastic Modulus – The higher the elastic modulus of the fibers
relative to that of the matrix, the greater is the proportion of
load carried by the fibers in the composite prior to cracking,
and the less the composite strain after the matrix has cracked
when the fibers carry all of the load.

Elasticity – Fibers which are not truly elastic and instead are prone to creep at normal or elevated temperatures tend to suffer stress relaxation in a loaded composite prior to cracking and time-dependent strain after cracking, either of which reduces their reinforcing effectiveness.

Poisson's Ratio – If the Poisson's ratio of the fiber material is significantly greater than the 0.20–0.25 applicable to most cementitious matrices, there is a tendency under tensile stress for the fiber to contract laterally more than the matrix and tear away from it, thus adversely affecting the adhesive bond (Fig. 3.1, right) (Kelly, 1975). Fibers with low modulus and high Poisson's ratio, for example polypropylene and nylon, may be prone to this tendency. However, shrinkage of the matrix surrounding the fiber probably has a mitigating effect on this tendency which is far from being fully understood (Kelly, 1975).

Most of the fibers considered for use in cementitious matrices have tensile strengths and elongation characteristics far superior to the normal matrix values, have Poisson's ratios comparable to the matrix, and low creep tendencies at normal temperatures, the exceptions being polymeric fibers like polypropylene and nylon. The most important difference between them from the point of view of reinforcing effectiveness is with respect to elastic modulus and creep behaviour. Truly elastic, high-modulus fibers offer greater potential reinforcing effectiveness than low-modulus elastic fibers or creep-prone fibers where the modulus decreases with loading time. However, realization of full reinforcing potential depends strongly on the interfacial shear bond between fiber and matrix and whether composite failure ultimately occurs by fiber pullout or fiber breakage.

3.1.2. Interfacial Shear Strength

In the simplest case of very long fibers aligned in the direction of uniaxial tensile stress (Fig. 3.2, upper left), just like conven-

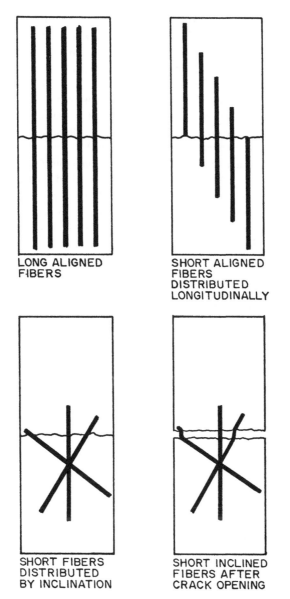

FIGURE 3.2 Hypothetical arrangements for long aligned fibers (upper left), short aligned fibers distributed longitudinally (upper right), short randomly oriented fibers (lower left), inclined fiber pullout (lower right)

tional straight reinforcing bars, it should be obvious that if
adhesive interfacial shear bond does not exist no tensile stress
can develop in the fibers. In this case the strength of the composite is the same as the strength of the matrix because the
fibers pull out of the matrix without resistance. This of course
neglects frictional shear when the Poisson's ratio of the fiber is
less than that of the matrix and shear due to mechanical interlock produced by surface texturing (like deformed rebar) or
improved end anchorage (like bent rebar).

In contrast, when there is very strong interfacial shear,
whether by adhesion, friction or mechanical interlock, the
fibers become subject to the entire load carried by the composite once the matrix cracks, and the ultimate strength depends
only on the amount and intrinsic strength of the fibers.
The critical fiber volume fraction, $V_{f(cr)}$, for this case is the
minimum amount needed to support the load just after the
matrix cracks, and is a function of the fiber strength, σ_{fu}, fiber
modulus, E_f, and the matrix strength, σ_{mu}, and cracking strain,
ε_{mu} (Hannant, 1978).

$$V_{f(cr)} = \frac{\sigma_{mu}}{(\sigma_{fu} - \varepsilon_{mu}E_f + \sigma_{mu})}$$

Hannant uses this simplistic analysis given in most texts on
composite materials to obtain values of $V_{f(cr)}$ of 0.3%, 0.4%
and 0.8% for steel, glass and polypropylene fibers respectively.
However, he notes that the assumptions on which it is based,
namely linear elastic behaviour of fibers and matrix, brittle
matrix behaviour to cracking, Poisson's ratio of zero for both
fibers and matrix, fibers aligned in the loading direction, and
complete bond between fiber and matrix, are unrealistic for
most practical fiber-reinforced cementitious composites. For
real composites several other factors considerably complicate
the relatively simple case of long aligned fibers (Fig. 3.2,
upper left).

3.1.2.1. *Pullout Versus Fracture for Short Aligned Fibers*

Even a strongly bonded fiber may pull out from the matrix before breaking if its actual embedded length relative to the plane at which cracking develops is short enough. Assuming for the moment that the crack occurs halfway along the fiber length (embedded length $\ell/2$), there is a pullout load and corresponding maximum fiber stress at the crack, σ_{fp}, less than the fiber strength σ_{fu}, that depends on embedded fiber length. It follows that as fiber length increases there is a critical fiber length, ℓ_c, below which fibers pullout of the matrix and above which they break instead of pulling out of the matrix. This conceptual dependence of fiber pullout load on embedded fiber length is confirmed experimentally for the specific case of steel fibers in mortar from single aligned fiber pullout tests (Fig. 3.3) (Gray, 1982). The fiber pullout load is actually

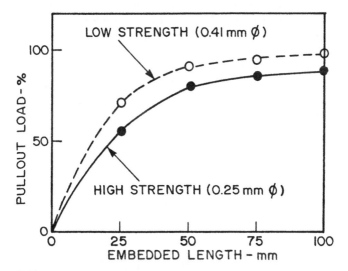

FIGURE 3.3   Pullout load as percentage of wire breaking load vs. embedded fiber length for uniform straight smooth steel wire (Gray, 1982)

governed by the contact area with the matrix which depends not only on embedded length, $\ell/2$, but on the circumference of the fiber cross-section, πd in the simplest case of a circular fiber. Thus, the pullout load is proportional to $\pi d\ \ell/2$, and the derived fiber stress is proportional to $(\pi d\ \ell/2)/(\pi d^2/4)$ which means that the fiber stress, σ_{fp}, is proportional to the fiber aspect ratio, ℓ/d, with an upper limit $\sigma_{fp} \leq \sigma_{fu}$ for aspect ratios $\ell/d \leq \ell_c/d$. (Fig. 3.4) (Broutman and Krock, 1967).

3.1.2.2. Variability of Embedded Fiber Length

In reality, lengthwise distribution of multiple aligned fibers relative to a crack in the matrix means that the embedded fiber length is not half the fiber length as assumed in the previous section. Instead, for multiple aligned fibers the shortest embedded portion of the length varies from zero to the fiber half length according to the fiber location relative to the crack (Fig. 3.2, upper right). This suggests an average fiber embedment length of $\ell/4$ for uniformly distributed fibers (Aveston et al., 1974) (Hannant, 1978). It also means that even when all the fibers are of length equal to the critical length, ℓ_c, about half of them fail to develop a stress equal to the fiber strength and their reinforcing effectiveness is reduced. Accordingly, it follows that the greater the fiber length and aspect ratio the greater the proportion of fibers likely to maximize their re-

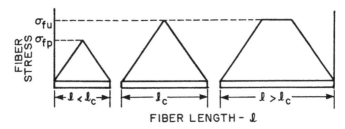

FIGURE 3.4   Critical fiber length vs. fiber stress (Broutman and Krock, 1967)

inforcing effectiveness by developing a tensile stress, σ_{fp}, equal to the fiber strength, σ_{fu}.

3.1.3. Effect of Fiber Orientation

Varying the orientation of fibers with a typical embedment length, for example $\ell/4$ when aligned in the direction of applied load, causes the actual embedded length to decrease as the fibers become more oblique to the crack (Fig. 3.2, lower left). Also, the number of fibers intersecting the plane of the crack decreases. Consequently, the reinforcing effectiveness of a group of fibers with 2-dimensional orientation varying systematically from $+90°$ to $-90°$ from the direction of applied stress (Fig. 3.2, lower) is expected to be less than that of the same group of fibers aligned parallel to the applied stress (Fig. 3.2, upper right), at least for the smooth straight fibers normally considered. Theories abound on the derivation of efficiency factors to reflect fiber reinforcing effectiveness in hypothetical 3-dimensional and 2-dimensional random fiber orientations, but these are rarely supported by confirmatory experimental data (Hannant, 1978).

Experimental fiber pullout tests in which the orientation of the fiber is varied in a controlled manner provide some insight into the behaviour of oblique fibers in real composites. The results from different investigations are most easily compared in terms of the pullout load of an inclined fiber relative to the pullout load for the corresponding aligned fiber. Both Gray (1982) and Naaman and Shah (1975 and 1976) employed nearly identical testing techniques using pairs of symmetrically arranged smooth straight steel fibers at various inclinations to the loading axis. Gray's (1982) results for a 50 mm embedded fiber length show a distinct increase in the relative peak fiber load at the time the matrix cracks with increase in fiber inclination (Fig. 3.5, upper), while Naaman and Shah's (1976) results for a 13 mm embedded length vary with fiber diameter. When both sets of results are considered in terms of

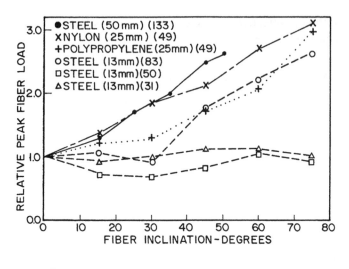

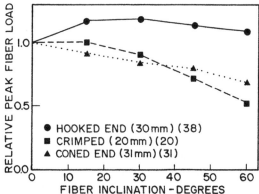

FIGURE 3.5   Relative peak fiber pullout load vs. fiber inclination for smooth straight fibers (upper) (Gray, 1982) (Naaman and Shah, 1976) (Li, Wang and Backer, 1990) and deformed steel fibers (lower) (Banthia and Trottier, 1994). Fiber length and embedment aspect ratio given in parenthesis.

fiber embedment aspect ratio, it seems clear that for smooth rigid steel monofilaments the effect of fiber inclination on peak fiber load is negligible at low fiber embedment aspect ratios and becomes increasingly significant at higher fiber embedment aspect ratios. This is consistent with the importance of fiber aspect ratio previously discussed. For comparatively more flexible nylon and polypropylene fibers with an embedded length of 25 mm Li, Wang and Backer (1990) also report an increase in peak fiber load with increasing fiber inclination (Fig. 3.5, upper) similar to the trend for steel fibers of higher embedment aspect ratio.

In contrast to the increase in relative pullout load with increasing fiber inclination observed for smooth straight fibers (Fig. 3.5, upper), the results of Banthia and Trottier (1994) for hooked-end, crimped, and coned-end fibers show somewhat different trends (Fig. 3.5, lower). For the smooth hooked-end fibers with low fiber embedment aspect ratio, the effect of fiber inclination is slight, as for the low aspect ratio fibers of Naaman and Shah (1976), so the hooked end appears to have little effect. However, for the crimped and coned-end fibers with much higher pullout loads for the aligned fiber orientation, there is an apparent decrease in relative peak fiber load with increasing fiber inclination. Overall, the results seem to indicate that the greater the fiber embedment aspect ratio the greater is the relative increase in pullout resistance attributable to fiber inclination for smooth straight fibers where adhesion is the primary factor governing pullout of the aligned fiber. For fibers with enhanced mechanical anchorage sufficient to cause mechanical interlock to override the adhesion-associated effect of aspect ratio, the comparatively higher pullout resistance for the aligned fiber neutralizes or reverses the effect of fiber inclination.

To further clarify the possible effects of fiber inclination it is instructive to compare the average interfacial shear stresses, τ_{ib}, calculated as pullout load divided by embedded interfacial area for the various steel fibers investigated (Table 3.1).

TABLE 3.1 Interfacial shear bond stress, τ_{ib}, vs. fiber inclination effect for steel fiber

Data source	Embedment aspect ratio	Profile	τ_{ib} — MPa	Inclination effect
Gray (1982)	133	Smooth and straight	1.13	Significant, +ve
Naaman and Shah (1976)	83	Smooth and straight	1.02	Significant, +ve
Naaman and Shah (1976)	50	Smooth and straight	2.62	Neutral
Naaman and Shah (1976)	31	Smooth and straight	2.62	Neutral
Banthia and Trottier (1994)	38	Smooth and straight	3.79	Neutral
Banthia and Trottier (1994)	20	Crimped	10.76	Significant, −ve
Banthia and Trottier (1994)	31	Coned-end	9.87	Significant, −ve

This shows that when τ_{ib} is very low fiber inclination improves it, as in the case of the higher fiber embedment aspect ratios of Gray (1982) and Naaman and Shah (1976) (Fig. 3.5, upper). When τ_{ib} increases for smooth straight fibers, as in the case of the lower aspect ratio fibers of Naaman and Shah (1976) and the hooked-end fiber of Banthia and Trottier (1994), the fiber inclination effect is neutralized. Finally, when τ_{ib} increases significantly as a result of enhanced mechanical interlock the fiber inclination effect is reversed, as for the crimped and coned-end fibers (Fig. 3.5, lower).

Post-peak fiber load also depends on fiber inclination. For the smooth straight steel fibers tested by Naaman and Shah (1976) the final load varies from zero to a significant percentage of the peak value for fiber inclinations of 60–75° (Fig. 3.6), particularly for the largest fiber diameter. For such highly inclined fibers several factors singly or in combination may help to increase pullout resistance. They include the effort needed for fiber bending at both extremities of the crack space as the crack widens (Fig. 3.2, lower right), and the greater role

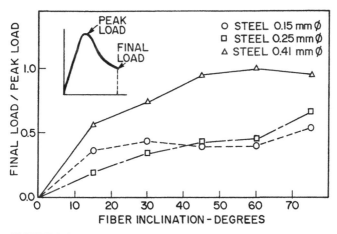

FIGURE 3.6   Relationship between peak and final pullout loads for smooth straight steel fibers (Naaman and Shah, 1976)

of friction compared with adhesion particularly where the fiber is pressed against the matrix at its entry into the crack space. Li, Wang and Backer (1990) calculated snubbing friction coefficients to distinguish the differences in performance between nylon and polypropylene fibers (Fig. 3.5, upper). The snubbing effect is expected to be most significant for fibers with greatest resistance to bending. This is supported by the fact that the final load is most affected by fiber inclination for the largest steel fiber diameter in Fig. 3.6.

The energy or work needed to pull a fiber completely out of the matrix reflects toughness, and is represented by the area under the load-slippage relationship. High final pullout loads sustained to high fiber slippages represent high pullout energy and greater toughening effectiveness. Pullout energy relative to the aligned fiber condition increases markedly with fiber inclination for the smooth straight relatively rigid steel fibers evaluated by Gray (1982) and Naaman and Shah (1976), and also for the smooth straight relatively flexible nylon and polypropylene fibers tested by Li, Wang and Backer (1990) (Fig. 3.7, upper).

Once again, the results of Banthia and Trottier (1994) show somewhat different trends for deformed strongly anchored fibers where the pullout energy is comparatively higher for the aligned fiber condition (Fig. 3.7, lower) than it is for the smooth straight fibers (Fig. 3.7, upper). Accordingly, there is a decrease in relative pullout energy with increase in fiber inclination, with the crimped and coned-end fibers showing the most marked decline (Fig. 3.7, lower). For these fibers the pullout form of failure that is normal for smooth monofilaments can be replaced by failure due to either wedge or core fracture of the matrix, or fiber breakage in tension, which markedly reduces the final slippage and the energy represented by the area under the load-slippage relationship (Fig. 3.8). The coned-end fiber was shown to be particularly prone to fiber failure and inferior pullout energy in a high strength matrix, even for the aligned fiber condition, demonstrating that

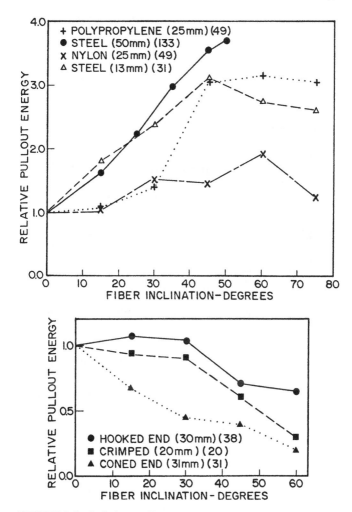

FIGURE 3.7 Relative pullout energy vs. fiber inclination for smooth straight fibers (upper) (Gray, 1982) (Naaman and Shah, 1976) and deformed steel fibers (lower) (Banthia and Trottier, 1994). Fiber length and embedment aspect ratio given in parenthesis.

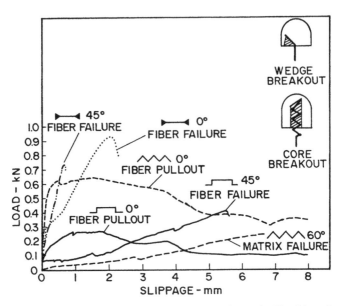

FIGURE 3.8   Failure modes in a high-strength matrix (Banthia and
Trottier, 1994)

extremely strong fiber anchorage is not necessarily optimal
(Fig. 3.8) (Banthia and Trottier, 1994).

In summary, the effect of fiber orientation as depicted by
single fiber pullout tests on aligned and inclined fibers is
complex and variable depending on factors such as fiber
embedment aspect ratio, fiber profile and end anchorage
enhancement, fiber tensile strength and matrix tensile and
shear strength. For smooth straight monofilaments such as
steel, nylon and polypropylene, random fiber orientation is
apparently beneficial to composite strength and toughness
because of the intrinsically low pullout resistance of these
fibers when aligned with the applied load. For some deformed
fibers with an enhanced profile or end anchorage, random fiber
orientation is apparently not beneficial because of the much
higher pullout resistance of such fibers when aligned, and the

possibility for inclined fibers of matrix fracture or fiber fracture replacing pullout as the dominant form of failure.

3.1.4. Group Fiber Interaction

The possibility exists that the per fiber contribution to reinforcing effectiveness for multiple fibers may be different from the contribution indicated from single fiber pullout tests. Here again, the pullout results for aligned and oblique fiber groups tested by Naaman and Shah (1976) may be relevant to understanding real composite behaviour. For aligned fibers no significant differences in the per fiber contribution to pullout load (Fig. 3.9) or pullout energy were observed for fiber groups up to the equivalent of a fiber content of 3% by volume. However, for oblique fiber groups inclined at 60° to the applied load, the per fiber contribution to both pullout load (Fig. 3.9) and pullout energy (Fig. 3.10) decreased significantly with increase in fiber content for smooth straight fibers. This may be attributable to the extent of wedge or core breakout in the matrix being more severe due to group fiber action at higher fiber contents. Apparently, the reduction in pullout energy for

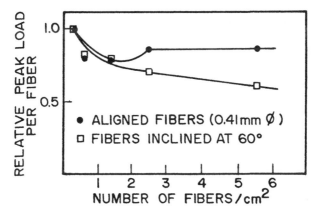

FIGURE 3.9   Pullout load per fiber for inclined fiber groups (Naaman and Shah, 1976)

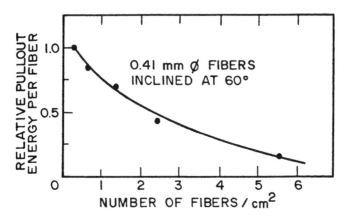

FIGURE 3.10   Pullout work per fiber for inclined fiber groups
(Naaman and Shah, 1976)

oblique fibers due to group fiber interaction (Fig. 3.10) tends
to offset any increase in single fiber pullout energy for oblique
fiber-matrix combinations (Fig. 3.7, upper), at least for the
smooth straight fibers evaluated.

3.2. OPTIMAL FIBER DEVELOPMENT

Consideration of the factors identified in the previous section
shows that there are fiber parameters which can be altered in a
controlled manner to beneficially influence fiber performance,
and that there are other parameters which are largely uncon-
trollable or unavoidable in practice. The variability of embedment length and number of fibers intersecting a unit area with
the distribution and orientation of the fibers and any effect due
to group fiber interaction are examples of largely uncontrol-
lable factors. However, the intrinsic geometrical characteristics of the fiber and the nature of the interfacial bond between
fiber and matrix are alterable and controllable to some extent.
Also, the intrinsic material properties of a fiber, such as tensile

strength, elongation and elastic modulus can in some cases be altered by processes involving heat or mechanical work.

3.2.1. Fiber Profile

Innovations that combine real reductions in fiber length or aspect ratio, which minimize the adverse effects of high aspect ratio on consistency and workability in the freshly mixed state previously mentioned, with increases in fiber pullout resistance, which improve reinforcing effectiveness in the hardened state, have been emerging for many years. Straight smooth monofilaments of uniform cross-section are now less common. Monofilament fibers with a lengthwise profile that is crimped or twisted rather than straight, or ends which are hooked or enlarged in various spaded (2-dimensional) or conical (3-dimensional) shapes are now common (Fig. 3.11). The notion of designing the ends or profile of steel fibers to increase pullout resistance and raise the peak fiber stress closer to the fiber strength has been explored for many years, starting with the early patents for steel fibers in the form of rings in 1931

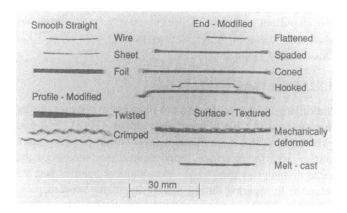

FIGURE 3.11   Monofilament fiber profiles — Uniform, enlarged-end, hooked-end, spaded end, coned-end, crimped, paper clip, circular

Fiber-Reinforced Cements and Concretes

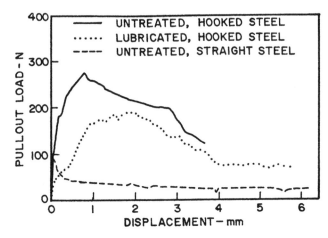

FIGURE 3.12   Comparative pullout resistance of straight and hooked steel fibers (Naaman and Najm, 1991)

(Naaman, 1985) and continuing into the 1990's with fibers incorporating cone-shaped ends (Destrée and Sahloul, 1991) or a paperclip profile (Rossi and Chanvillard, 1992). The improved pullout resistance of steel fibers with hooked ends has been demonstrated in single fiber pullout tests (Fig. 3.12) (Naaman and Najm, 1991).

3.2.2. Fiber Form

Multifilament strands that are not intended to separate during mixing, typically 204 filament glass strands (Fig. 3.13) are an alternative to the monofilament form which provides more surface area for interfacial bond than monofilaments of the same diameter as the strand. Multifilament bundles that are intended to separate during mixing, typically bundled polyester (Fig. 3.13), greatly increase the number of fibers and the interfacial bond area per unit volume of concrete. Steel fiber bundles held together with water-soluble glue that allows separation during mixing are widely available (Fig. 3.13). Partial

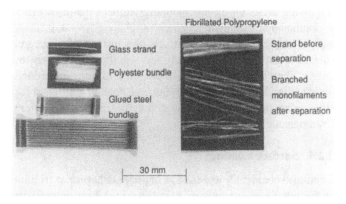

FIGURE 3.13   Multifilament strands, bundles and fibrillated tapes

splitting (fibrillation) to produce multifilament mesh-like strands with secondary crosslinks that are intended to separate into branched monofilaments during mixing with improved mechanical bond due to the branching is characteristic of fibrillated polypropylene (Fig. 3.13).

3.2.3. Surface Deformation and Texture

Surface deformation can improve the components of pullout resistance due to friction and mechanical interlock for any type of fiber, and has been used mainly to alter the performance of straight uniform fibers without special end anchorage, such as steel (ACI Committee 544, 1996), polyethylene (Kobayashi and Cho, 1981) and nylon (Li, Wang and Backer, 1990). Surface texturing during the manufacturing process can roughen the surface to improve frictional resistance or adhesive bond. Steel and stainless steel fibers cast directly from the liquid metal are an example of roughening to improve mainly the frictional bond between fiber and matrix (De Guillebon and Sohm, 1986). The embedment of silica fume particles in the surface of otherwise smooth poorly adhesive polypropylene is an example of frictional roughening combined with

improving adhesion by incorporating particles that establish a
hydraulic bond with hydrating cement (Esbjerg Tovvaerkstabrik
A/S, 1986). Examples of the effects of a surface lubricant and
surface deformation, separately and together, are shown for
single fiber pullout using nylon with a 25 mm embedded
length (Fig. 3.14, upper) (Li, Wang and Backer, 1990) and for
steel with the same embedded length (Fig. 3.14, lower)
(Naaman and Najm, 1991).

3.2.4. Surface Coatings

Coatings specifically intended to improve adhesion or friction
at the interface, or coatings intended primarily to protect the
parent fiber from damage during mixing or chemical attack by
the constituents of cement or chemical admixtures, may
influence reinforcing effectiveness by altering fiber pullout
resistance. Such coatings have been of particular concern for
glass fibers prone to attack by the alkalis in cement (Bijen,
1986) (Bentur and Diamond, 1987) (Hayashi, Sato and Fujii,
1986). A chemical surface treatment of PAN-type carbon
fibers to improve adhesive bond with cement by boiling with a
solution of sodium chlorate and sulphuric acid, washing, and
immersing in γ-aminopropyl triethoxysilane solution has been
shown to improve composite properties such as tensile and
flexural strength (Chen, Cheng and Gao, 1991).

3.2.5. Material Processing

Annealing and cold-working processes alter the yield and ultimate tensile strengths, elongation, and bending stiffness of
steel fibers. Annealing an initially cold-rolled crimped steel
fiber for 15 minutes at 600°C reduced the peak fiber pullout
load by about 40% and caused a corresponding drop in fiber
hardness and by implication yield strength (Banthia et al.,
1992). Chemical additives and mechanical working alter the
strength, elongation and elastic modulus of some polymeric
fiber materials. Stretching and heat treatment have been used

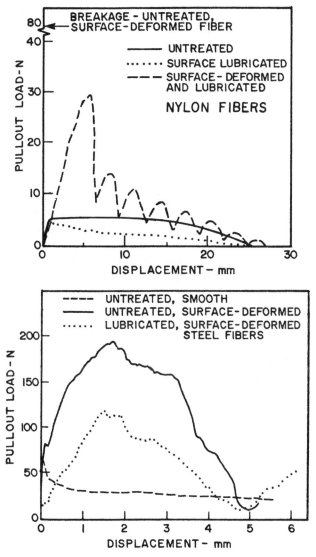

FIGURE 3.14   Comparative pullout resistance of smooth and surface-
deformed nylon fibers (upper) (Li, Wang and Backer,
1990) and steel fibers (lower) (Naaman and Najm,
1991)

to increase the elastic modulus and reduce the elongation of polypropylene, and were combined with surface treatment to improve adhesive bond with cement paste (Krenchel and Jensen, 1980).

3.2.6. Fiber Size

Absolute fiber size determines the number of fibers per unit of batched weight and the number per cubic meter of matrix. Since the total weight rather than the absolute size reflects the material cost of the fibers, the question arises whether a large number of small fibers offers better reinforcing effectiveness than the same weight of a smaller number of large fibers with the same aspect ratio. For example, in one testing program (Johnston and Skarendahl, 1992) involving five commercially available but not geometrically similar types of steel fiber, the number of fibers ranged from 4000 to 100,000 per kilogram. Intuitively, the highest number of fibers per cubic meter of matrix might be expected to be optimal because of their close spacing and consequent ability to inhibit propagation of small microcracks into larger macrocracks. However, there is some very limited evidence for steel fibers indicating that, regardless of aspect ratio, the fiber length should exceed the maximum aggregate size in the matrix (Blood, 1970) (Tatro, 1985). This criterion has tended to become a broadly accepted "rule of thumb", although there is little actual data to support it. Thus, very small short fibers, sometimes termed microfibers, which are long enough to be effective in cement paste or mortar matrices may be less effective in concrete than the same weight of larger longer fibers having the same aspect ratio, surface texture and end anchorage. Fibers currently marketed for use in concrete are rarely shorter than 15 mm or larger than 65 mm, but can be as short as 1–3 mm in cement paste matrices.

3.2.7. Effect of Matrix Characteristics

Changing matrix characteristics such as water-cement ratio and aggregate-cement ratio affects the adhesive component of interfacial shear bond strength (Gray and Johnston, 1987). An example is the influence of water-cement ratio and associated matrix strength on the pullout resistance of hooked steel fibers, where pullout load and pullout energy increase with increasing matrix strength (Fig. 3.15) (Naaman and Najm, 1991). These authors also observed changes in pullout performance with the presence of admixtures. Latex had a significant effect while silica fume and fly ash had only minor effects.

3.3. COMPATIBILITY OF FIBER AND MATRIX

Fibers must be physically and chemically compatible with the moist alkaline environment characteristic of cementitious matrices. Certain fibers may be attacked either by water or by the alkalis present in solution in cement paste. For example, many cellulose-based vegetable fibers such as cotton, sisal and hemp rot in a moist environment, and glass fibers made from

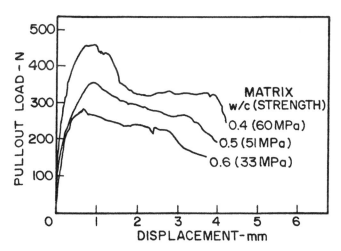

FIGURE 3.15   Comparative pullout resistance of hooked steel fibers in different matrices with sand-cement ratio 2.0 (Naaman and Najm, 1991)

E-glass and some types of polyester fiber are prone to deterioration on contact with cement alkalis. Fiber-specific compatibility problems are discussed subsequently for each type of fiber.

In general, the possibility of fiber deterioration from contact with the constituents of cement and chemical or mineral admixtures should always be considered, especially when there is little or no documentary evidence confirming satisfactory long-term resistance to such deterioration. ASTM Standard C 1116 (ASTM, 1994) for fiber-reinforced concrete and shotcrete has a requirement that evidence of satisfactory long-term performance be provided when there is doubt about fiber-matrix compatibility.

Even when the parent fiber material is known to be compatible with all ingredients in the concrete, problems can arise with coatings used intentionally to facilitate fiber manufacture or introduced unintentionally as contaminants with the raw material or manufacturing process. A recent unusual case involved fibers sold as steel fibers (where compatibility with concrete is normally not an issue) that were prepared using some aluminium-coated raw material. This resulted in a confusing scenario where some batches and test cylinders behaved normally while others exhibited abnormal cracking and very low compressive strengths.

CHAPTER 4

Composite Behaviour Under Load

Composite behaviour is the result of the various fiber-matrix interactions discussed in the preceding chapters for the freshly mixed and hardened states. In practice, the main concern is with short and long-term behaviour after hardening, but it must be emphasized that obtaining composites likely to perform satisfactorily in the hardened state depends on dealing properly with the fiber-matrix interactions in the freshly mixed state discussed previously. This means ensuring that the selected manufacturing process results in combination of the prescribed matrix with the specified type and amount of fibers to produce a properly consolidated composite with the fibers uniformly distributed and undamaged by the manufacturing process. Any combination of poor consolidation, fiber clumping or balling, fiber damage during mixing, or chemical incompatibility of fiber and matrix can ruin the composite performance of individually excellent raw materials.

Once again, consideration of behaviour under uniform uniaxial stress in tension and compression facilitates initial understanding of how composite performance is quantified. This initial understanding can then be extended to more complex loading conditions such as shear and bending that are often of greater interest in practical applications.

49

4.1. DIRECT TENSION

To establish the potential upper bound levels of performance possible in fiber-reinforced cementitious composites, it is useful to first consider cement paste matrices with preplaced high volume fractions of long aligned fibers when interfacial bond is high and fiber pullout is not an issue (Fig. 4.1.) (Majumdar and Laws, 1991). This case offers the greatest possible reinforcing effectiveness in terms of maximizing improvements in strength and toughness at each fiber content. In contrast, practical composites with discontinuous randomly oriented relatively short fibers of the same type, glass in this case, have a comparatively smaller maximum fiber content

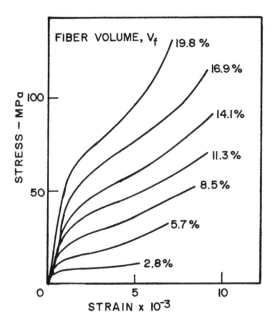

FIGURE 4.1   Tensile stress-strain curves for continuous aligned glass fibers in cement paste (Majumdar and Laws, 1991)

limited by mixing considerations, and the fibers may be short
enough for fiber pullout to influence composite failure (Fig.
4.2) (Majumdar and Laws, 1991). While the initial portions of
the stress-strain curves are similar in shape at comparable fiber
volume fractions (3–8%) in Fig. 4.1 and 4.2, the breaking
stresses are significantly lower for the composites with randomly oriented shorter fibers, reflecting the decreased reinforcing effectiveness attributable to shorter fiber length with
consequently lower aspect ratio, and probably to a lesser
extent 2-d random fiber orientation. Important similarities are
that as the fiber volume fraction increases there is an improvement in the stress at what is termed the first crack or limit of
proportionality (LOP), identified by the sharp decrease in the
slope of the stress-strain curve associated with the loss of composite stiffness caused by the onset of cracking in the matrix.
There are also corresponding improvements in breaking
strength, ductility as represented by strain at failure, and

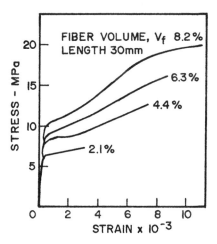

FIGURE 4.2   Tensile stress-strain curves for 30 mm long randomly
oriented glass fibers in cement paste (Majumdar and
Laws, 1991)

energy absorption capability or toughness as represented by
the total area under the stress-strain curve (Fig. 4.2).

Just as fiber content is the dominant factor influencing the
improvements in first-crack strength (strengthening) and in
energy absorption to failure (toughening) imparted by randomly oriented discontinuous glass fibers (Fig. 4.2), its effect
is similar for carbon fibers in a medium strength cement paste
(Fig. 4.3, left) (Akihama, Suenaga and Nakagawa, 1988).
However, increasing the matrix strength by reducing the
water-cement ratio causes loss of post-crack toughness, probably attributable to a stronger interfacial resistance to fiber
pullout that allowed fiber fracture rather than pullout to domi-
nate the composite behaviour (Fig. 4.3, right). The secondary
but significant importance of fiber aspect ratio, which corresponds to length for multifilament glass strands, is illustrated
by the increase in ductility and toughness associated with
increased fiber pullout resistance at longer fiber lengths (Fig.
4.4) (Majumdar and Laws, 1991). In general, the stress-strain
curves in Fig. 4.3 and 4.4 show significant improvement in

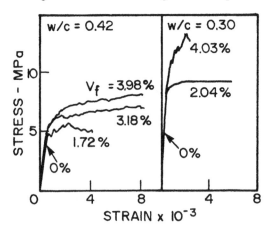

FIGURE 4.3   Tensile stress-strain curves for 10 mm long randomly
             oriented carbon fibers in cement paste (Akihama,
             Suenaga and Nakagawa, 1988)

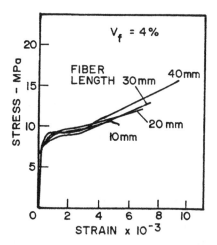

FIGURE 4.4   Effect of strand length (aspect ratio) on tensile stress-
strain behaviour for glass fibers in cement paste
(Majumdar and Laws, 1991)

first-crack strength and substantial toughness thereafter with a
positive slope to ultimate sometimes characterized as strain-
hardening behaviour, except when fiber breakage rather than
pullout dominates and toughness is adversely affected (Fig.
4.3, right).

For conventionally mixed mortars or concretes, workability
and mixing considerations previously mentioned limit fiber
contents to a maximum of 1–2%, much smaller than the 2–8%
range represented in Fig. 4.2–4.4. For such composites the
fibers are typically present in amounts less than the critical
volume fraction, and composite failure by fiber pullout occurs
almost immediately after the matrix cracks. For example, in
the case of 0.6–1.8% by volume of steel fiber in a mortar
matrix increases in strength and strain at failure are quite small
in a normal (low stiffness) tension test, but still reflect the individual influence of fiber content and fiber aspect ratio (Fig.
4.5, left) (Johnston and Coleman, 1974). The combined
significance of fiber aspect ratio and fiber content represented

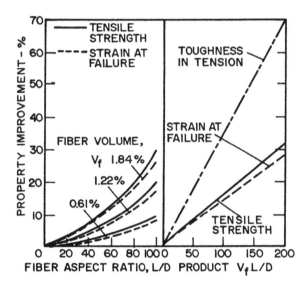

FIGURE 4.5   Influence of fiber content and aspect ratio on compos-
              ite properties individually (left) and in combination
              (right) for steel fiber-reinforced mortars in tension
              (Johnston, 1980)

by their product V_f L/D is also apparent in terms of its
influence on strength, ductility and toughness (Fig. 4.5, right)
(Johnston, 1980). Likewise, the effect of increasing the fiber
pullout resistance of smooth fibers by altering the mortar proportions to increase the average interfacial shear bond strength
by about 40%, or by replacing smooth fibers with rougher
surface-deformed equivalents, reflects the increase in composite strength expected from the previous discussion of single
fiber pullout tests (Fig. 4.6) (Gray and Johnston, 1987).

In conventional tension tests at relatively low fiber contents
the stiffness of the testing equipment is usually so low that
failure occurs once the matrix cracks (Fig. 4.5 and 4.6). When
the testing equipment is artificially stiffened, or is inherently
very stiff, a portion of the stress-strain curve following the first

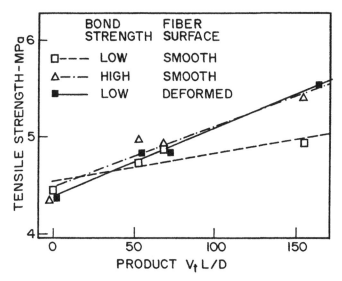

FIGURE 4.6   Effects of increased fiber pullout resistance on relationships between tensile strength and V_f L/D for
steel fiber-reinforced mortars (Gray and Johnston,
1987)

crack can be detected, making the effects of changes in fiber
parameters more obvious. For example, in the case of smooth
straight steel fibers post-crack toughness represented by the
area under the curve after first crack is clearly influenced by
fiber content (Fig. 4.7) (Gopalaratnam and Shah, 1987) and by
changes to the fiber profile that improve fiber anchorage and
resistance to pullout, such as hooked or enlarged ends
(Fig. 4.8) (Shah et al., 1978). Various types of synthetic fibers
exhibit a similar sensitivity to fiber content as well as to characteristics such as fiber modulus and interfacial bond strength
(Fig. 4.9) (Wang, Li and Backer, 1990). In general, for these
comparatively small fiber volume fractions below the critical
$V_{f(cr)}$, the stress-strain curves show either a gradual decrease in
stress following the first crack, sometimes termed strain-
softening behaviour (Fig. 4.8 for hooked or enlarged-end

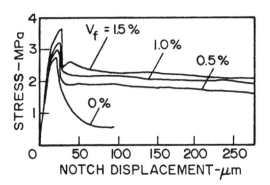

FIGURE 4.7   Tensile stress-strain curves for different fiber contents
in steel fiber-reinforced mortar (Gopalaratnam and
Shah, 1987)

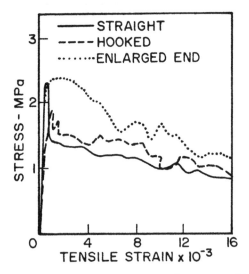

FIGURE 4.8   Tensile stress-strain curves for mortar with 1.73%
volume of different types of steel fiber (Shah et al.,
1978)

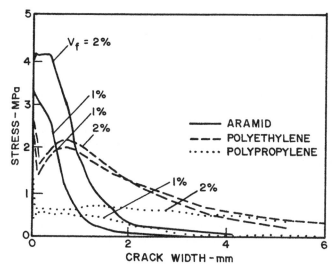

FIGURE 4.9   Tensile stress-strain curves for mortars reinforced
with synthetic fibers (Wang et al., 1990)

fibers), or in varying degrees a sudden decrease in stress after
first crack followed by further strain development at near-con-
stant load for smooth fibers without improved end anchorage
(Fig. 4.9). Whether tests using artificial stiffening are relevant
in practice depends on the nature of the application. They
appear irrelevant for a load-controlled situation, where cata-
strophic failure occurs when the stress reaches the matrix
strength. They are more relevant to a strain-controlled situation where stress relaxation can occur and reduced stress can
be sustained by the fibers after the peak stress. However, the
implied post-peak strength for the mortar without fibers (0% in
Fig. 4.7) seems meaningless.

At the higher fiber contents and aspect ratios possible using
the slurry infiltration process, tensile strengths are much
higher and the effects of increasing fiber aspect ratio and
improving interfacial bond are magnified. In the case of the
silica-flour cement mortars reinforced with 10% by volume of

relatively rough melt-extract steel fibers increasing the aspect ratio and improving interfacial bond by autoclave curing substantially increase tensile strength (Fig. 4.10) (Baggott and Abdel-Monem, 1992).

The preceding discussion has tended to highlight the influence of fiber characteristics on composite performance. However, interfacial shear strength is naturally influenced by the matrix as well as the fiber. While changing matrix proportions to increase interfacial shear strength may merely raise a relatively low fiber stress to a higher value still well below the fiber strength and improve composite performance accordingly (Fig. 4.6), changing them sufficiently when the interfacial bond is already relatively high may cause the fiber stress to reach the fiber strength and the failure to change from ductile, predominantly fiber pullout, to brittle, predominantly fiber

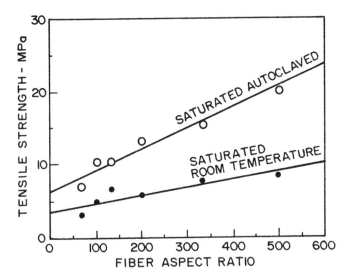

FIGURE 4.10   Effect of increasing aspect ratio on tensile strength of slurry-infiltrated steel fiber reinforced cement paste (Baggott and Abdel-Monem, 1992)

fracture, as shown for carbon fibers (Fig. 4.3). Brittle composite failure can also be promoted by using high strength cementitious matrices which are inherently more brittle with respect
to cracking or splitting caused by breakdown of the mechanical interlock between fiber and matrix. Changing matrix
characteristics like cement type, water-cement ratio, fine
aggregate-cement ratio, and fine-coarse aggregate ratio, or
including additives like latex, silica fume or fly ash, have all
been shown in the previous chapter to affect the fiber pullout
resistance of steel fibers (Naaman and Najm, 1991), and can
be expected to influence composite performance to some
extent regardless of fiber type.

4.2. COMPRESSION

Compressive stress does not tend to cause cracking of the
matrix, at least not directly in a plane perpendicular to the
stress as happens for tension, so fibers are expected to have a
negligible reinforcing effect in the direction of applied stress.
However, the uniaxial compressive test normally used to evaluate behaviour in compression produces a combination of
shear failure near the ends of the specimen (conical or pyrami-
dal fracture) with lateral swelling of the unconfined central
section accompanied by cracking parallel to the loading axis
when the lateral strain exceeds the matrix cracking strain in
tension. Fibers can affect these facets of uniaxial compressive
behaviour that involve shear stress and tensile strain. This is
evident in the post-peak portion of the stress-strain curve as
increased strain capacity and correspondingly increased toughness represented by the area under the curve, but for low fiber
volume fractions ($\leq 2\%$) it is only obtainable for inherently
stiff or artificially stiffened testing systems (Fig. 4.11) (Fanella
and Naaman, 1985). Once again, whether this is relevant in
practice or not depends on the nature of the application and the
anticipated terminal serviceability condition. For example,

under extreme conditions, such as earthquake loading, fibers may assist in reducing the degree of disintegration and loss of structural integrity associated with large strains in compression.

Under normal (low stiffness) testing conditions failure in compression occurs at or soon after the peak load with little post-peak toughness for composites of low fiber content (≤ 2%). In general, the fibers have little effect on compressive strength calculated from the peak load, and both slight increases and decreases in strength have been reported with increasing fiber content. The decreases are probably attributable largely to incomplete consolidation detectable by changes in density (Johnston and Skarendahl, 1992), since consolidation becomes more difficult with increasing fiber content and aspect ratio, as discussed in chapter 2. A direct comparison of fiber-reinforced composite and unreinforced matrix strengths confirms that the effect of steel fibers on strengthening is minimal for conventionally mixed concretes (Fig. 4.12) (Johnston, 1994–1).

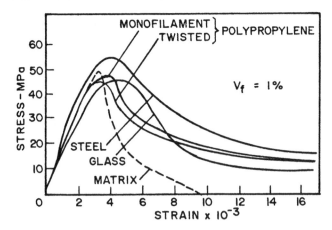

FIGURE 4.11   Compressive stress-strain curves for fiber-reinforced mortars with various types of fiber (Fanella and Naaman, 1985)

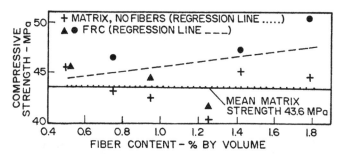

FIGURE 4.12   Effect of fibers on compressive strength for two
types of steel fiber in concrete (Johnston, 1994–1)

In contrast, for high fiber contents (> 10%) such as are
achievable with the slurry infiltration process, fibers can
improve performance significantly, probably because the mode
of failure is shear on a plane at about 45 ± 15° to the loading
axis. However, the improvements depend on the orientation of
the fibers (primarily in horizontal plane during slurry infiltration) relative to the axis of loading. Both compressive strength
(based on peak load) and toughness (based on total area under
the curve) are generally highest when the fibers are primarily
in the plane of lateral tensile strain perpendicular to the
applied compressive load (Fig. 4.13) (Homrich and Naaman,
1987). For high fiber content thin-section composites prepared
by the spray or premix processes performance in compression
is rarely reported, the implication being that it is of little interest or relevance in practice.

4.3. BENDING

Composite behaviour in bending reflects the dominant
influence of the extreme fiber stress in the tension zone relative to the cracking stress. However, since the stress and strain
distributions are no longer uniform, and are only determinable
and meaningful prior to the onset of cracking in the matrix,

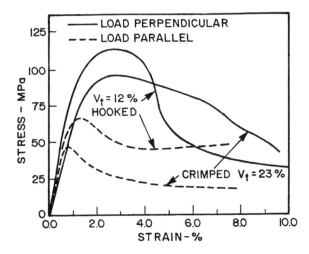

FIGURE 4.13   Compressive stress-strain curve for slurry-infiltrated
                cement pastes with steel fibers (Homrich and
                Naaman, 1987)

behaviour in bending is usually examined in terms of load and
deflection rather than stress and strain.

4.3.1. First-Crack Flexural Strength

Prior to the onset of cracking in the matrix, cementitious
matrices are usually assumed to behave elastically with consequent linear distributions of stress and strain symmetrical
about a neutral axis at the half-depth of the beam, so the standard elastic bending formula is applicable for determining the
maximum flexural stress. The upper limit for validity of this
formula is reached at the onset of cracking, and the corresponding stress calculated according to the loading arrangement, span and size of beam is the first-crack flexural strength
(Fig. 4.14).

For low fiber content composites increasing the interfacial
shear strength and fiber pullout resistance by changing matrix
proportions or replacing smooth steel fibers with surface-

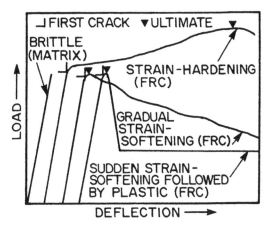

FIGURE 4.14   Typical load-deflection curves for matrix and fiber-reinforced composites (FRC) in bending

deformed equivalents has effects on first-crack flexural strength (Fig. 4.15) (Gray and Johnston, 1987) similar to those previously identified for direct tension (Fig. 4.6). For higher fiber content composites made with steel or carbon fibers, or hybrid combinations of each, fiber characteristics such as type and amount affect first-crack flexural strength even more noticeably in terms of the greater degree of strengthening achievable (Fig. 4.16) (Banthia and Sheng, 1991).

4.3.2. Post-Crack Behaviour

Despite the fact that composite behaviour is best assessed in terms of load and deflection because the formula for elastic bending is not applicable to a cracked beam, there is an unfortunate and widespread tendency to discuss performance after cracking in terms of a fictitious stress corresponding to the ultimate flexural strength or modulus of rupture derived using the maximum load in the formula (Fig. 4.14). With regard to serviceability in terms of the severity of deflection and cracking, even the maximum load has little significance because it

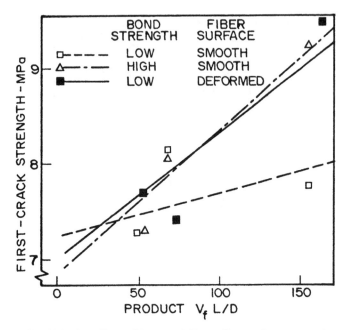

FIGURE 4.15   Effects of increased fiber pullout resistance on rela-
                tionships between first-crack flexural strength and V_f
                L/D for steel fiber-reinforced mortars (Gray and
                Johnston, 1987)

can occur either immediately after first crack (strain-softening
behaviour) or long after first crack (strain-hardening behaviour) (Fig. 4.14). Much more important are the levels of
residual load capacity and toughness (area under the load-
deflection curve) that remain at various deflections after
cracking.

Tests using specimens instrumented with strain gauges at
various depths show that after the onset of cracking in the
matrix the neutral axis moves gradually from the half-depth
towards the compression face as load increases (Fig. 4.17,
upper) (Akihama, Suenaga and Banno, 1986–1). It can reach
as little as 20% of the beam depth for steel fiber-reinforced

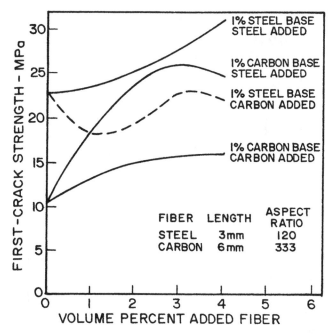

FIGURE 4.16   Effects of fiber type and fiber content on first-crack
flexural strength of cement pastes reinforced with
steel and carbon fibers (Banthia and Sheng, 1991)

concrete (Edgington, 1973) and glass fiber-reinforced cement
(Allen, 1971). Strain measurements for carbon fiber-reinforced
cements suggest a value of d_n/D of 25% near ultimate load for
thin specimens, increasing slightly for thicker specimens
(Fig. 4.17, lower) (Akihama, Suenaga and Banno, 1986–1).
While the strain distribution is essentially linear, various forms
of nonlinear stress distribution can prevail in the tension zone
(Hannant, 1978) (Fig. 4.18) depending on whether the volume
of fibers and their interfacial shear resistance is sufficient to
avoid pullout when the matrix first cracks ($V_f > V_{f\ (critical)}$), or
not ($V_f < V_{f\ (critical)}$), corresponding respectively to strain-
hardening or strain-softening behaviour (Fig. 4.14). One form

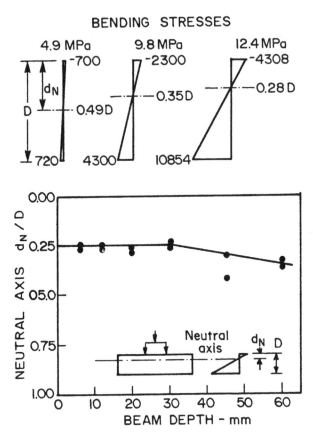

FIGURE 4.17   Change in strain distribution and neutral axis as load
               increases for 20 mm thick section (upper), and effect
               of section thickness on neutral axis position (lower)
               for carbon fiber-reinforced cement (Akihama,
               Suenaga and Banno, 1986–1)

of idealized distribution (Hannant, 1978) is to assume the
neutral axis at 25% of the beam depth with elastic distribution
of stress in the compression zone and plastic distribution of
stress in the tension zone (Fig. 4.19, right). When compared

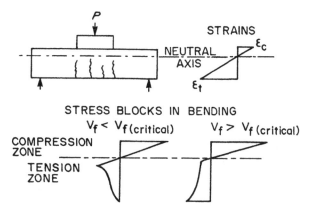

STRESS BLOCKS IN BENDING

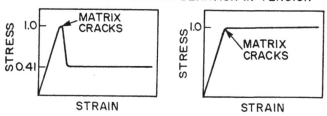

RELATIVE STRESS-STRAIN BEHAVIOR IN TENSION

FIGURE 4.18   Possible bending stress blocks (upper) and corresponding stress-strain patterns in tension (lower) for strain-hardening and strain-softening composites (Hannant, 1978)

with the elastic distribution of stress throughout the depth that approximates the stress distribution for an unreinforced matrix (Fig. 4.19, left), it is apparent that the upward shift of the neutral axis enlarges the relatively weak tension zone while reducing the much stronger underutilized compression zone. This explains in terms of maximum bending moments why improvements in performance attributable to fibers are greater for bending than for direct tension. For example, a post-crack tensile strength, f_t, of only 0.41 times the first-crack tensile

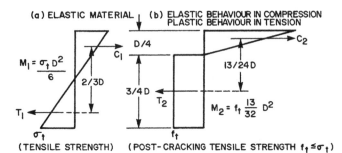

(TENSILE STRENGTH)   (POST-CRACKING TENSILE STRENGTH $f_t \leq \sigma_t$)

FIGURE 4.19   Flexural stress distributions and moment capacities
for idealized elastic brittle matrix (left) and composite with plastic stress distribution in the tension zone
(right) (Hannant, 1978)

strength (Fig. 4.18), σ_t, is sufficient to sustain flexural load
capacity at the first-crack load value (M_1 would be equal to M_2
in Fig. 4.19), while a post-crack tensile strength equal to σ_t
increases moment capacity to 2.4 times the first-crack value
($M_2 = 2.4\,M_1$ in Fig. 4.19).

Naturally, what actually happens quantitatively for any particular fiber-matrix combination depends on how fiber type,
content and interfacial shear strength alter the shape of the
stress blocks and the position of the neutral axis idealized in
Fig. 4.18 and 4.19. For example, in the high fiber content
carbon fiber-reinforced cements with 10 mm long pitch-based
fibers, where bending stresses after first crack reflect the magnitude of the load but do not of course represent true values of
stress in the composite, there is modest strengthening (compared with the unreinforced matrix) up to the first crack followed by increased residual strength and toughness after
cracking sustained to considerable deflection for the weaker
matrix (Fig. 4.20, left) (Akihama, Suenaga and Banno,
1986–1). In contrast, strengthening to first crack is greater for
the stronger matrix, but post-crack residual strength and
toughness are not sustained to as high a deflection (Fig. 4.20,
right). The difference probably reflects a change from rela-

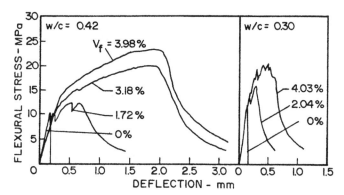

FIGURE 4.20   Behaviour of carbon fiber-reinforced cements with
10 mm long fibers in two matrices of different
strength (Akihama, Suenaga and Banno, 1986–1)

tively low fiber-matrix interfacial bond strength with failure
primarily by fiber pullout to failure dominated by fiber breakage associated with relatively high interfacial bond. This data
source also provides a rare opportunity to contrast the reinforcing effectiveness of these fibers in bending (Fig. 4.20) with their
effectiveness in direct tension (Fig. 4.3) for specimens of comparable thickness and fiber orientation, and confirms the considerably higher stress levels attainable in flexure, particularly
after cracking, analytically predicted from Fig. 4.19. Further
evidence of the dependence of the nature of the failure in the
composite on matrix characteristics and their associated effect
on the interfacial bond strength is seen in the relative toughness
levels achieved with different fiber-matrix combinations
(Fig. 4.21) (Akihama, Suenaga and Banno, 1986–1). Failure is
primarily brittle with the highest strength paste matrix and more
ductile for the lower strength paste and mortar matrices.

For lower fiber content composites like steel fiber-
reinforced concrete, there is again a wide range of behaviour
that reflects the influence of fiber and matrix characteristics.
Generally, strengthening to first crack is not very significant at
fiber volume fractions less than 1% (Fig. 4.22) (Johnston,

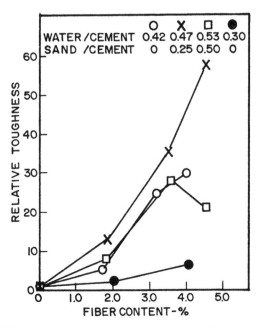

FIGURE 4.21   Relative toughness of carbon fiber composites
with different cement paste and mortar matrices
(Akihama, Suenaga and Banno, 1986–1)

1994–1), and actual values mainly reflect the strength of the
parent matrix as determined by water-cement ratio and density
achieved after consolidation (Fig. 4.23) (Johnston and
Skarendahl, 1992). The most significant effects are in terms of
residual strength and toughness after first crack. Superior
residual strength and toughness sustained to considerable
deflections are associated with the higher fiber pullout resistance produced by high aspect ratio and hooked-end anchorage, and lower residual strength and toughness are associated
with straight fairly smooth low-aspect-ratio fibers (Fig.
4.24)(Johnston, 1991) (Johnston and Skarendahl, 1992). These
differences are evident regardless of fiber content and matrix
strength, and demonstrate the dominant importance of fiber
profile and aspect ratio.

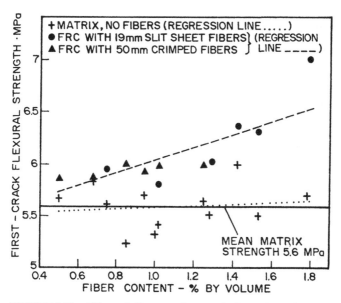

FIGURE 4.22   Effect of fibers on first-crack flexural strength for two types of steel fiber in concrete (Johnston, 1994–1)

The problem of how to quantify degrees of strengthening or toughening in bending in a way that might be applicable to design, specification compliance and quality control is a continuing challenge. Engineers who have no specialist expertise with fiber-reinforced cements or concretes have difficulty understanding the significance of toughness in relation to structural performance, and a preference has developed for the use of residual strength as the criterion for judging post-crack composite performance because its significance is more obvious. For fiber-reinforced concretes, ASTM Standard C 1018 (ASTM, 1994) addresses the problem by defining parameters, such as I_{20} and $R_{10,20}$ (Fig. 4.24), that represent respectively toughness to a specified end-point deflection and residual load averaged over a specified deflection interval expressed as a percentage of the first-crack load. The tough-

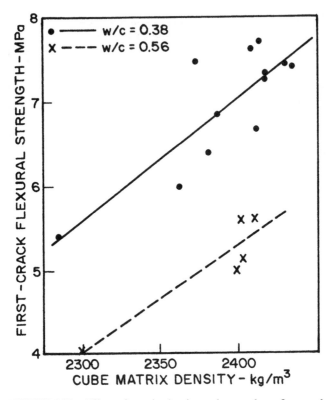

FIGURE 4.23   Effect of matrix density and strength on first-crack flexural strength of fiber-reinforced concretes with several types of steel fiber (Johnston and Skarendahl, 1992)

ness index I_{20} is the ratio of the area up to the specified end-point-deflection (10.5 times the first-crack deflection) divided by the area up to the first-crack deflection. The residual strength factor $R_{10,20}$ applies to the deflection interval from 5.5 to 10.5 times the first-crack deflection. For ease of understanding, and as a reference for level of performance, it should be recognized that a load-deflection relationship showing perfectly elastic-plastic behaviour, typical of mild steel, corre-

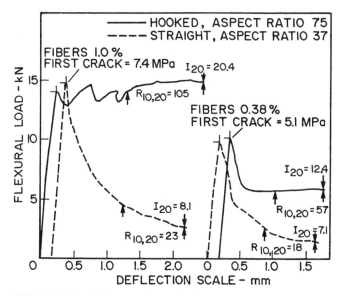

FIGURE 4.24 Differences in residual strength and toughness for high-aspect-ratio (75) hooked-end fibers and low-aspect-ratio (37) straight steel fibers in a strong concrete matrix (left) and a weaker concrete matrix (right) (Johnston, 1991)

sponds to $I_{20} = 20$ and $R_{10,20} = 100$. The curve for 1% of 75 aspect ratio hooked fibers in the 7.4 MPa matrix approximates this behavior, while the other curves depict inferior performance (Fig. 4.24). Flexural performance evaluation using ASTM C 1018 criteria and other proposed alternatives is discussed further in Chapter 7.

4.4. SHEAR

Failure by delamination can be the critical form of shear failure in thin-section high fiber content composites fabricated by a layering process that causes a high degree of

2-dimensional fiber alignment in the plane of the section with few fibers crossing layer boundaries. For example, in short beam tests on glass fiber-reinforced cements with primarily 2-dimensional fiber orientation, interlaminar shear strength parallel to the fibers was no higher than for the parent matrix (Oakley and Unsworth, 1978). In contrast, shear strength perpendicular to the plane of the section is much improved by the fibers, and is approximately equal to the in-plane direct tensile strength (Fig. 4.25) (Oakley and Unsworth, 1978). A similar weakness in shear on planes parallel to the fibers has also been reported for slurry-infiltrated steel fiber-reinforced mortar with a fiber content of the order of 10% by volume (van Mier and Timmers, 1992).

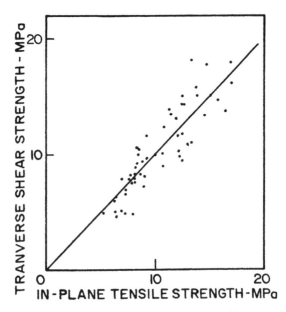

FIGURE 4.25   Transverse shear strength vs. tensile strength for thin-section glass fiber-reinforced cements (Oakley and Unsworth, 1978)

For conventionally mixed thick-section composites with primarily 3-dimensional random fiber orientation, shearing of the matrix without involvement of the fibers is very unlikely. Improvements in shear strength, ductility and toughness attributable to fibers are best assessed in torsion tests which produce pure shear, rather than in beam tests which combine shear and bending. Several sources of data are reviewed by Nanni (1990), and they show both strengthening (based on maximum torque) and toughening (based on area under the torque-twist relationship), for example in steel fiber concretes (Narayan and Kareem-Palanjian, 1983) (Fig. 4.26). Shear strength increases up to 80% are reported for 2% by volume of steel fibers of aspect ratio 75 (Tegos, 1989). Improvements in torsional shear strength correspond closely with those for direct tension (Edgington, Hannant and Williams, 1974). This

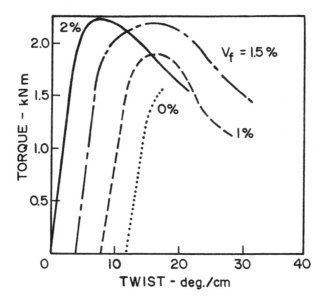

FIGURE 4.26   Torque-twist relationships for steel fiber-reinforced concrete (Narayan and Kareem-Palanjian, 1983)

equality between shear and tensile strengths reflects the role of
principal tensile stress as the criterion governing shear failure
in pure torsion and diagonal tension failure in short-span deep
beams where delamination is not an issue. Improvements in
torsional toughness tend to reflect the influence of fiber
content and fiber aspect ratio (Fig. 4.27) (Craig et al., 1986),
as for bending or direct tension.

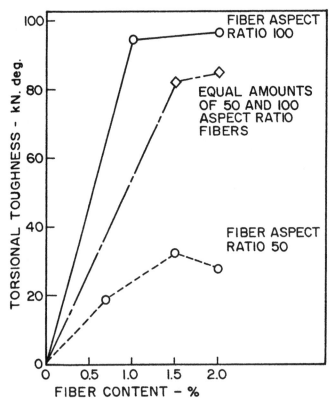

FIGURE 4.27   Torsional toughness of steel fiber-reinforced concrete
(Craig et al., 1986)

4.5. EFFECTS OF PREFERENTIAL FIBER ALIGNMENT

Preferential fiber alignment no matter how caused can affect composite behaviour under all modes of loading, although most of the experimental data supporting this view relate to bending. Aligned 1-dimensional fiber orientation arrangements produced by special techniques not readily applicable in practice confirm superior performance in bending attributable to fiber alignment in the direction of applied load (Bergstrom, 1975) (Hannant, 1974) (Fig. 4.28). However, the principle has been incorporated into some practical applications discussed in Chapter 7 (Rotondo and Weiner, 1986) (Skarendahl, 1992).

4.5.1. Thin Fiber-Reinforced Cement Sheets

Preferential 2-dimensional fiber orientation enforced by the boundary surfaces in thin-section products is common in fiber-reinforced cements. Strength and toughness parallel to the

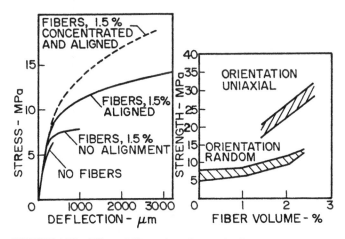

FIGURE 4.28   Effect of fiber orientation on the flexural performance of steel fiber-reinforced concretes (Bergstrom, 1975) (Hannant, 1974)

plane of the section are enhanced at the expense of these properties in a direction perpendicular to the plane of the section which are generally unimportant. However, the fibers provide minimal reinforcing against delamination by shear parallel to the plane of the section. Typical results for carbon fiber-reinforced cement with 4% by volume of 10 mm long fibers in sections 10 to 60 mm thick show that the effect of fiber alignment becomes quite significant when the ratio of specimen thickness to fiber length drops below 3.0 (Fig. 4.29) (Akihama, Suenaga and Banno, 1986–1).

Obviously, the effect of fiber alignment can be expected to increase with decrease in the ratio of specimen thickness to fiber length. In glass fiber-reinforced cements using fiber lengths of 25–50 mm this ratio is often less than 0.5, so the effect of fiber alignment is considerable, possibly increasing the strength by 50–70% when the ratio is 0.5 (Fig. 4.29).

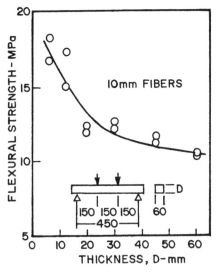

FIGURE 4.29   Effect of section thickness on flexural strength of carbon fiber-reinforced cements (w/c = 0.42) (Akihama, Suenaga and Banno, 1986–1)

4.5.2. Fiber-Reinforced Concrete

In thin sections some degree of 2-dimensional fiber orientation in the plane of the section may be promoted not only by the boundary surfaces but by processes such as the impact of the fibers against a hard surface in shotcreting or by surface finishing with a vibrating screed. For thick sections where fiber orientation in the interior of the composite remote from the boundary surfaces is normally 3-dimensional random, intense external vibration is reported to cause a tendency for fiber alignment in a horizontal plane (Hannant, 1972) (Swamy and Stavrides, 1975). However, the boundary surfaces, whether formed or finished, always promote alignment of the fibers in their vicinity. Its extent certainly depends on the length of the fibers and possibly also on their flexibility, as flexible fibers may have less tendency to align with the boundary surface than rigid fibers.

The effect of fiber alignment for relatively long (75 mm) and relatively short (25 mm) rigid steel fibers was examined by comparing standard molded beams in which fiber alignment by the mold surfaces is maximized with beams of the same 100 mm square final size after sawing to remove a zone of highest fiber alignment near the boundary surfaces, 25 mm thick along each side and 50 mm thick along the bottom. (Fig. 4.30) (Johnston, 1989). For the short fiber, the 25 mm thickness removed by sawing (Fig. 4.30) is equal to the fiber length, so the orientation of the short fibers within the sawn boundaries is likely to be essentially random and unaffected by the original boundary surface.

Considering first-crack strength, the similarity between strengths for molded and sawn specimens with 25 mm fibers confirms that these fibers in a 100 mm square molded specimen achieve an essentially random orientation. For the longer 75 mm fibers there is obviously more alignment in a 100 mm square molded specimen than in sawn specimens originally 150 mm square, which may explain the lower strength for the sawn specimens (Fig. 4.30).

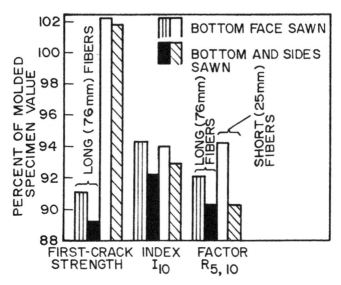

FIGURE 4.30   Effect of removal by sawing of the boundary surfaces containing aligned fibers on flexural performance of steel fiber-reinforced concrete (Johnston,
1989)

Considering ASTM C 1018 toughness and residual strength
parameters, I_{10} and $R_{5,10}$, shows that removal of aligned fibers
of any length near the boundary surface reduces these parameters by 8–10% compared with equivalent molded specimens where the surfacial fibers are not removed (Fig. 4.30).
The results also suggest that the presence of aligned fibers
near the bottom boundary surface of the beam, where stress
and strain are highest, is particularly critical for post-crack
performance which is primarily fiber-dependent. Aligned
fibers at the side boundary surfaces, where the stress and strain
are lower, are less important, since their removal by sawing
has little additional effect over that of removing the fibers near
the bottom surface.

Applying the same logic to standard molded specimens, it follows that using small specimens tends to enhance the measured performance. Comparing 150 mm square and 100 mm square beams with 63 mm rigid steel fibers (ratios of minimum specimen dimension to fiber length of 2.4 and 1.6 respectively) showed that the first-crack flexural strength was only slightly increased, while toughness and residual strength parameters were significantly increased by using the smaller specimen in which the fiber alignment is more pronounced. The increases attributable to using the smaller specimen were 6% for first-crack strength, 20% for the I_{20} toughness index, and 29% for the residual strength factor $R_{10,20}$ (Johnston, 1989). Again, this is consistent with first-crack strength being primarily matrix-dependent and post-crack performance being primarily fiber-dependent.

4.5.3. Consensus and Consequences

The consensus of the results depicted in Fig. 4.29 and 4.30 is that fiber alignment by boundary surfaces tends to increase first-crack strength and toughness in bending when the ratio of specimen thickness or minimum cross-sectional dimension to fiber length drops below a critical value of 3.0 to 4.0. The resulting increases in post-crack toughness and residual strength are greater than for first-crack strength because they are primarily fiber-dependent.

Some important practical implications of fiber alignment are that thin small test specimens tend to exhibit properties superior to the thicker structural sections they are supposed to represent and to specimens sawn from such sections. However, the converse is also possible when molded specimens are thicker than the section in the actual application, as often happens with bridge deck overlays or shotcrete linings. Accordingly, test specimens should reflect practical circumstances as much as possible with regard to preparation process and geometrical characteristics, particularly thickness in relation to fiber length. ASTM C 1018 (ASTM, 1994) contains relevant provisions for fiber-reinforced concrete, and ASTM C 947 (ASTM, 1995) for glass fiber-reinforced cements. Whether fiber flexibility mitigates the fiber alignment effect attributable to formed and finished boundary surfaces is unclear. However, both the carbon fibers represented in Fig. 4.29, which are quite flexible, and the steel fibers represented in Fig. 4.30, which are rather rigid and unlikely to bend, are in terms of flexural properties subject to the effects of fiber alignment by boundary surfaces.

4.6. CONTROLLED FIBER DISTRIBUTION

Concentrating fibers in regions of maximum stress rather than distributing them uniformly throughout a structural unit is not only more efficient in a technical sense, but can offer significant cost saving where the use of a fiber-reinforced composite in conjunction with an unreinforced matrix is practical. Sprayed or shotcreted applications finished with a final layer of unreinforced matrix can overcome difficulties associated with fibers at an exposed surface where fiber protrusion or corrosion are undesirable. Beams or slabs, including pavements, built using monolithic sandwich construction with the fiber-reinforced composite only in the portion of the depth subject to substantial tensile stress can be equivalent in performance to full-depth fiber-reinforced sections. Confinement of fibers to a relatively thin layer also tends to promote preferential fiber alignment as discussed previously, so sandwich construction can combine the benefits of both fiber alignment and reduced fiber consumption.

Evaluation of the principle of sandwich construction in bending, initially by (Rahimi and Kesler, 1979), and later by (Rahimi and Cao, 1990), showed that beams with a varying thickness of steel fiber-reinforced mortar in the tension zone failed in a brittle manner until the thickness of the fiber-reinforced layer reached about 17% of the depth (Fig. 4.31). For fiber-reinforced layer thicknesses from 17% to 50% of the depth, the ASTM C 1018 toughness indices I_5, I_{10} and I_{30} all improved with increasing layer thickness, as exemplified by the I_{30} index (Fig. 4.31), and the beams failed in a ductile manner. Flexural strengths for the 17% sandwich were up to 30% higher than for full-depth fiber-reinforced beams, but did not improve further with increasing layer thickness. For this particular fiber-matrix combination, with 18 mm fibers in a 100 mm deep beam, reinforcing only the bottom 35 to 40% of the beam depth appears optimal from the point of view of strength and toughness (Fig. 4.31). In this case, the thickness of the fibrous layer is about twice the fiber length. Clearly, longer fibers would undergo a greater degree of alignment and should offer even better performance in terms of the minimum layer thickness needed to produce significant strengthening and toughening in sandwich construction.

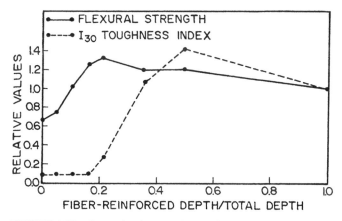

FIGURE 4.31   Strengthening and toughening in sandwich-constructed steel fiber-reinforced mortar beams (Rahimi and Cao, 1990)

CHAPTER 5

Materials, Mixture Proportioning and Manufacture

The nature of the fibers and the cementitious matrix strongly influence the mixture proportioning, mixing and fabrication procedures appropriate for any particular composite because of the wide range of possible fiber-matrix interactions described in Chapters 2 and 3. Successful manufacture depends on selecting a manufacturing process, fiber type, fiber amount and matrix proportions that will achieve uniform dispersion of the fibers throughout the chosen matrix without placement and consolidation problems associated with the mixture-stiffening effect of the fibers and without physical damage to the fibers caused by the manufacturing process.

5.1. FIBER-REINFORCED CEMENTS AND MORTARS

Despite being commonly identified as fiber-reinforced cements, the matrix in these composites, although primarily cement-based, normally contains significant amounts of fine sand, and sometimes supplementary cementing materials such as fly ash, blastfurnace slag, or silica fume. The volume fraction of the paste phase is relatively high compared with concrete matrices, and consequently so is the maximum fiber amount possible without undue mixture stiffening, as shown previously (Fig. 2.3 and 2.6). Fiber-reinforced cements are

85

often fabricated into thin-section components where the section thickness is less than the fiber length, so the fibers are consequently aligned to a significant degree in the plane of the section which improves their reinforcing effectiveness. Historically, asbestos-cement, first produced at the end of the 19th century, has been the most widely used form of thin-section fiber-reinforced cement, but it is gradually being supplanted by cements reinforced with glass, carbon and other fibers free of the health hazard associated with asbestos.

5.1.1. Glass Fiber-Reinforced Cement

Following the recognition that borosilicate E-glass fibers originally developed in the 1950's for use in plastics were subject to rapid deterioration caused by the alkalis in cement, considerable effort was directed towards developing an alkali-resistant (AR) type of glass fiber. This led to the development of a zirconia-based glass fiber produced in the form of a bonded multifilament strand initially marketed under the name Cem-FIL in the 1970's (Majumdar and Laws, 1991).

Currently, the material known as glass fiber-reinforced concrete (GFRC) in N. America, or as glass fiber-reinforced cement (GRC) in Europe, is fabricated in two ways, both of which employ multifilament glass strand using a glass with a minimum of 16% zirconia. Each strand comprises 50, 102, 204 or 408 individual 13–20 μm diameter filaments bound together by a non-dispersible coating material called a sizing that is intended to prevent the separation of the filaments and protect them from abrasion (PCI Committee, 1991). The sizing may incorporate a chemical inhibitor to reduce or delay the migration of calcium hydroxide into the strand which eventually bonds the individual filaments together thus embrittling the composite and adversely affecting long-term performance. A roving is a group of strands gathered together and wound into a continuous package. The roving, typically with 20–40 strands (Majumdar and Laws, 1991) (PCI Committee, 1993), normally separates into strands when chopped to the 13–50 mm lengths typically used as reinforcement in GFRC. The effectiveness of the sizing in preventing the separation of strand into filaments, called filamentising, and the extent to which the roving separates into strands during processing largely determine the surface area of the fiber reinforcement and its associated mixture-stiffening effect.

Additives such as polyethylene oxide or carboxyl methyl cellulose lubricate the fiber surface and increase the viscosity of the mixing water, thus helping to disperse the chopped strand and improve resistance to mixture segregation. They have also been particularly useful for mixing soft-coated strand that tends to filamentise easily (Ryder, 1975). However, a harder coating/sizing is better for maintaining strand integrity during mechanical mixing.

Naturally, minimizing or eliminating mechanical mixing from the manufacturing process helps to reduce filamentising and abrasive damage to the fiber reinforcement. This has resulted in two distinct manufacturing processes, premix which minimizes mechanical mixing and spray-up which eliminates mixing entirely.

5.1.1.1. Spray-Up GFRC

In the spray-up process, sometimes termed the spray method, the glass roving is fed through a chopper which cuts it to the required length and injects the chopped strands into the matrix slurry emitted through a spray gun. The sprayed slurry, atomized by compressed air, mixes with the chopped strand as it passes from the gun to the formwork surface against which successive layers of fiber-reinforced slurry are deposited manually or automatically (Fig. 5.1). Each layer is sprayed to a thickness of 3–6 mm and then compacted by rolling (Fig. 5.1, lower right) to fully wet the fibers, remove entrapped air and maximize density. A special ribbed roller that presses the fibers into the matrix is preferred to a flat roller that cannot

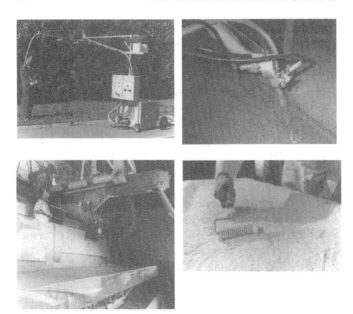

FIGURE 5.1   Spray-up placement process for GFRC — Manual
spraying (upper), automated spraying (lower, left),
roller compaction (lower, right). (Courtesy of Power-
Sprays Ltd., Building Research Establishment U.K.
and Nippon Electric Glass America Inc.

encapsulate the fibers into the matrix (Hanson et al., 1990).
The orientation of the spray gun to the sprayed surface and its
distance from the surface must be carefully controlled to
achieve uniformity in fiber orientation and distribution
(Hanson et al., 1990). Subsequent layers are sprayed and
rolled until the required thickness, normally 13 mm minimum,
is reached. Uniformity of thickness and fiber content are
important, particularly at raised corners where rolling tends to
remove material away from the corner. Scrim, a manufactured
AR glass fabric, is used to reinforce corners openings, anchorage zones, or other areas of panels subject to high localized
stress.

For architectural effect, the fiber-reinforced material, termed the backing mixture, is often preceded by a face mixture without fibers that may contain colouring pigments, sands or special aggregates selected specifically for appearance of the exposed panel surface. When a face mixture is used it must be as thin as practically possible, as the thickness and strength of the GFRC backing mixture subsequently sprayed on top of it can only be determined from the total combined thickness of both layers.

Spray-up GFRC mixtures using normal portland cement typically contain 5% of 25–50 mm long AR glass strand by weight of mixture. The minimum permissible fiber content is 4% (PCI Committee, 1991) (PCI Committee, 1993). Fiber contents more than 7% or strands longer than 50 mm lead to placement and compaction problems, while amounts less than 5% or shorter than 25 mm offer less reinforcing effectiveness, although placement and compaction are of course easier. For AR glass strand of specific gravity 2.7, 5% by weight corresponds to approximately 4% by volume of mixture.

The characteristics of the cement-based slurry fed to the spray gun are naturally important in determining mixture workability and roller-induced compactabilty at the formwork surface. Consistency is assessed by pouring the slurry into a 80 mm long, 57 mm diameter, plexiglass tube centred on a plexiglass base, gently rodding it to remove entrapped air, lifting the tube to allow the slurry to spread across a series of eight concentric rings 65–225 mm in diameter marked on the base, and recording the spread in terms of rings covered on a scale of 0 to 8 (PCI Committee, 1991). The consistency may vary from relatively stiff, zero-ring spread, to relatively fluid, 8-ring spread, depending on water-cement ratio, sand-cement ratio, and the presence of normal chemical admixtures (water-reducing, superplasticizing, accelerating, retarding etc.) or a thixotropic additive such as methyl cellulose used to retain mixture cohesion and reduce slump when spraying vertical surfaces. It also depends on whether or not a polymer additive is used to facilitate curing, since the water in the polymer additive becomes part of the mixing water and must be accounted for as part of the required water-cement ratio. Typical material requirements and mixture proportions for GFRC mixtures are given in Table 5.1 (PCI Committee, 1993) (PCI Committee, 1991). Normal, high early strength or other cements conforming to ASTM C 150 or equivalent are recommended. The color consistency of the cement can be very important, particularly for non-fiber facing mixtures or exposed GFRC mixtures.

Sand for the GFRC backing mixture should be a masonry mortar sand conforming to ASTM C 144 or equivalent with approximately 1 mm maximum size to minimize blocking of the spray nozzle. The sand gradation should be chosen to minimize the interparticle void content occupied by cement paste and therefore minimize shrinkage potential. The optimum gradation is affected by the shape of the sand particles, since the rounded particles typical of natural sands pack more tightly and pump more easily than the angular particles typical of manufactured sand. A single sand or a combination of sands

TABLE 5.1   Typical portland cement-based GFRC mixtures

Constituent	Spray-up	Premix
Chopped strand — %[a]	5[a]	3–3.5[b]
Strand length-mm	25–50	12–25
Sand/cement	1.0	0.75–1.0
Water/cement	0.30–0.35	0.30–0.35
Acrylic copolymer — %[c]	4–15	0–10
Defoamer — %[d]	–	0.2
Superplasticizer — %[d]	0–0.5	0–1.0

[a] Minimum 4% by weight (approximately 3.2% by volume) of mixture. Fibre specific gravity 2.7.
[b] Maximum 4% because of mixing and compaction considerations.
[c] Percent of polymer solids by weight of dry cement.
[d] Percent by weight of mixture. May be incorporated with the acrylic copolymer.

which when combined with the cement meets a Fuller gradation given by P = 100 (d/D)^n where n = 0.2 to 0.5 is recommended (PCI Committee, 1991). P is the total of the
percentages of sand and cement passing sieve size d for sand
of maximum particle size D. P for each sieve size is calculated
assuming that all of the cement passes the 75 μm sieve and all
of the sand is retained on this sieve. The value of n depends on
the proportions of sand and cement identified by the sand-
cement ratio. For example, P is 50% for the 75 μm sieve at a
sand-cement ratio of 1.0. This corresponds to a value of
n = 0.27 (Table 5.2) for the Fuller gradation to which the sand
should be matched. For sand-cement ratios of 1.5 and 2.0, the
values of P for the 75 m sieve are 40% and 33% respectively,
and the corresponding most suitable values of n are 0.35 and
0.43 respectively. The gradation of the matching sand, P$_s$, is
calculated from P$_s$ = (P – P (75 μm))/R, where R is the weight
fraction of sand in the mixture. Clearly, the ideal matching
sand gradation for sands of the correct maximum particle size
is only slightly affected by fairly large changes in the sand-
cement ratio.

Freshly mixed GFRC without a polymer additive requires
moist curing to promote cement hydration and prevent cracking due to shrinkage caused by loss of water. A minimum of 7
days at 95% or higher humidity to maintain a surface-moist
condition, and a temperature of at least 16°C are recommended (PCI Committee, 1993). Special care must be exer-
cised to prevent any drying during this curing period,
recognizing that thin-section components are more prone to
the effects of drying than more massive units with a lower
ratio of surface area to volume.

In view of the sensitivity of thin sections to drying and
cracking, an acrylic thermoplastic copolymer dispersion consisting of 45–55% polymer solids dispersed in water is often
used to facilitate curing of GFRC. The polymer creates a film
within the mixture and on the exposed surface which inhibits
escape of moisture and eliminates the need for moist curing.

TABLE 5.2   Fuller cement + sand gradations and matching sand gradations for different sand-cement ratios

Sieve size-μm	Sand/cement = 1.0 R = 0.50		Sand/cement = 1.5 R = 0.60		Sand/cement = 2.0 R = 0.67	
	Fuller P — % n = 0.27	Matching sand P$_s$ — %	Fuller P — % n = 0.35	Matching sand P$_s$ — %	Fuller P — % n = 0.43	Matching sand P$_s$ — %
1000	100	100	100	100	100	100
600	87	74	84	73	80	70
300	72	44	66	43	60	40
150	60	20	51	18[a]	44	16
75	50	0	40	0	33	0

[a] Percent sand at given sieve size, P$_s$, calculated as (51-40)/0.6 or P$_s$ = 11/0.6 = 18.

To ensure that the polymer dispersion coalesces into a film the curing temperature must exceed the minimum film formation temperature. The recommended minimum temperature is 16°C and the maximum should not exceed 50°C.

5.1.1.2. Premix GFRC

In the premix process the matrix ingredients are mixed together first using some form of mechanical mixer (Molloy, 1994) (PCI Committee, 1993, Appendix J). Then chopped AR glass strand is added gradually towards the end of the mixing process. Since the strands are exposed to greater potential damage by mechanical mixing than in the spray-up process, only high integrity chopped strand made specially for premix with a harder sizing than is used for spray-up should be employed. Strands for spray-up rovings have a softer sizing and are less resistant to separation into filaments, termed filamentising. The key to preventing damage to the strands is mixing them for the shortest possible time needed to achieve uniform strand dispersion and thorough wetting by the cement slurry, usually less than 2 minutes. Any evidence of tangling or filamentising indicates a mixing regime that is too long, or too severe, or both. One effective process employs a two-speed high shear mixer with the matrix premixed at the high speed and the strand added under low speed operation (Peter, 1994). Alternatively, a high shear mixer for the matrix and a separate premix mixer used during strand addition can be employed (Fig. 5.2, upper). Another effective alternative is the flexible-base Omni-mixer (Concrete Construction, 1986) in which mixing occurs by an undulating action of the base that turns the mixture over rather than shearing it by moving blades (Fig. 5.2, lower).

Like any mixing process that combines long slender fiber reinforcement with a relatively fluid matrix, the mixture-stiffening and workability-reducing effect limits the amount and aspect ratio or length of strand that can be incorporated even

FIGURE 5.2   Premix GFRC — Integrated high shear mixer, premix
mixer, weighing system and multi-roving glass fiber
chopper (upper) and flexible-base Omni-mixer (lower)
(Courtesy of Power-Sprays Ltd. and Nippon Electric
Glass America Inc.)

with the best mixers, and increases the risk of air entrapment and low strength caused by incomplete consolidation. Accordingly, fiber content is usually lower and strand length shorter than in the spray-up process (Table 5.1), so the reinforcing effectiveness is consequently less than in spray-up. A defoaming additive is often desirable to minimize entrapped air.

Nevertheless, provided proper consolidation is achieved, premix is useful for highly sculptured architectural units and flat or thick-section molded products where high strength is not the prime requirement.

Premix GFRC can also be sprayed using appropriate equipment with AR glass strand no longer than 20 mm (Molloy, 1994), but strengths are usually lower than for spray-up where the fiber content is normally greater, the strand length longer, and the strand alignment more favorable two-dimensionally in the plane of the section. Sprayed premix is not considered an alternative to spray-up for thin-section applications where high strength is important.

5.1.2. New GFRC Matrices

The recognition that GFRC made with normal portland cement tends to become brittle in the long term, even when using AR glass fibers, primarily due to calcium hydroxide produced during cement hydration being deposited between the fiber bundles, has led to the development of alternative matrix systems intended to alleviate this problem. They include the use of low-lime or lime-free cements to minimize calcium hydroxide generation, or normal cement combined with an active pozzolan to consume the calcium hydroxide as it is formed. Polymer additives have also been evaluated because they improve workability and placeability, facilitate curing as previously mentioned, and reduce moisture movement and shrinkage/expansion tendencies during drying/wetting (Majumdar and Laws, 1991).

Two new cementitious systems have been introduced in North America since 1993 (PCI Committee, 1993, Appendix

I). One uses a calcium sulphoaluminate, rapid hardening, hydraulic cement with an additive that contains plasticizer, set-retarder and a pozzolan to consume the calcium hydroxide produced during hydration (Molloy and Jones, 1993). Setting time depends on mixing temperature and is controlled by the retarding component in the additive. After setting, early hydration is strongly exothermic, usually within 30 minutes, and the composite must be water-cooled to ambient temperature. Subsequent hydration proceeds without the need for moist curing. The acrylic copolymer additives used for curing normal cement-based GFRC are not compatible with this cement.

The other system is based on enrichment of normal portland cement with 25% metakaolinite to consume the calcium hydroxide liberated during cement hydration (Thiery, Vautrin and Francois-Brazier, 1991) (van der Plas, Yue and Bijen, 1992). Unlike other pozzolans which react quite slowly with calcium hydroxide, metakaolinite is highly reactive and does not inhibit strength development at early ages. However, the high surface area of the metakaolinite increases water demand, and a superplasticizer is accordingly desirable to achieve satisfactory workability without an excessive water-cement ratio. Acrylic copolymer curing additives are optional to eliminate the need for moist curing.

Both matrix systems can be used in either the spray-up or premix manufacturing procedures. Recommended mixture proportions vary depending on the process (Molloy, Jones and Harmon, 1994) in much the same way as for conventional GFRC in Table 5.1, with fiber content and fiber lengths greater in the spray-up than in the premix process.

Many other alternatives to the normal portland-cement based matrix that include polymer additives or low-lime or lime-free cements, such as high alumina cement or supersulphated cement made from blastfurnace slag and calcium sulphate, have been investigated at the U.K. Building Research Station, (Majumdar and Laws, 1991). A low alkali, low shrinkage cement made with calcium silicates, calcium sulphoaluminate, anhydrite and blastfurnace slag, called CGC cement, has been used in Japan for manufacture of GFRC (Hayashi et al., 1992).

5.1.3. Carbon Fiber-Reinforced Cement

Carbon fiber-reinforced cements (CFRC) were initially investigated using relatively expensive PAN-type fibers produced from carbonizing polyacrylonitrile yarn. They are characterized by high modulus and high tensile strength. Subsequently, efforts to reduce the high cost of carbon fibers have led to development of a range of cheaper fibers produced from pitch that may, depending on the nature of the pitch, have either low or high modulus and low or high tensile strength.

Carbon fibers, unlike glass fibers, are not affected by the lime or alkalis in normal cement-based matrices, but, like glass and other relative fragile fibers, they are prone to damage by conventional mechanical mixing. The process of blending the fibers into the matrix is analogous to the process previously described for premix GFRC, and may utilize a special mixer such as an Omni-mixer to minimize damage to the fibers (Akihama et al., 1986–2) (Park and Lee, 1991), or a conventional mortar mixer with an appropriate mixing regime and dispersing additives (Ando et al., 1990) (Banthia, Moncef and Sheng, 1994).

Carbon fibers are generally characterized by high specific surface and high aspect ratio (Table 5.3), which makes them difficult to disperse uniformly in the matrix at fiber contents greater than about 1% by volume unless dispersing additives like carboxyl methyl cellulose and fine filler material are present. The matrix can be cement paste or a cement-filler paste with fly ash, fine silica sand, silica powder or silica fume. The finer the filler material, the better its dispersing effectiveness to the point that mixing can apparently be successful with a normal mortar mixer and 20% silica fume by weight of cement (Banthia, Moncef, and Sheng, 1994).

TABLE 5.3 Typical mixture characteristics for carbon fiber-reinforced cements prepared using an Omni-mixer or a normal mortar mixer

Source reference	Akihama et al. (1986–2)	Park & Lee (1991)	Ando et al. (1990)	Banthia Moncef and Sheng (1994)
Chopped fiber — %[a]	2–4.5	0.5–2.0	1–4	1–3
Length — mm	10	3–12	6–18	3–10
Diameter — μm	14.5	14.5	15–17	18
Aspect ratio	690	210–830	400–1060	170–560
Silica sand/cement				0.50
Silica powder/cement	0–0.5	0.4–0.5[c]		
Silica fume/cement		0.4[c]		0.2
Water/cement	0.30–0.53	0.30–0.53	0.45	0.35
Methyl cellulose — %[b]	1.0	0.5	0.25–1.0[d]	
Superplasticizer — %[b]	0–2	1–6	2.4	0.8–2.4
Defoaming agent — %[b]			0.1	

[a] Percent by volume of mixture. Fiber specific gravity 1.63.
[b] Percent by weight of cement.
[c] Either but not both.
[d] Depending on mixing regime.

Normally, a superplasticizing admixture is needed to ensure adequate mixture workability with a reasonably low water-cement ratio. Typical mixture characteristics are given in Table 5.3. The mixing regime may influence the need for a dispersing additive like carboxyl methyl cellulose. Ando et al. (1990) found that dry-mixing of the carbon fiber, cement, sand and dispersant prior to addition of water, superplasticizing admixture, and defoaming agent, reduced the amount of dispersing additive needed when compared with a mixing regime in which only part of the sand was added at the dry-mix stage. The reduced amount of dispersant and resulting low mixture viscosity also improved mixture workability and trowel finishability.

For any particular mixing regime the mixture-stiffening effect of the carbon fibers in terms of flow (determined using a standard mortar flow table) depends fundamentally on total fiber surface area per unit volume of mixture. This reflects the number of fibers in a unit volume of mixture and their diameter for any particular fiber length and amount expressed as a volume fraction of the composite. The thinner the fiber diameter, the greater the number of fibers of constant length in any given volume fraction and the higher the specific surface. Naturally, as the volume fraction increases the specific surface increases correspondingly, so this supports the notion of mixture-stiffening or reduction in flow as a function of fiber volume fraction and fiber diameter contained in the calculation of specific surface area of fibers per unit volume of mortar (Fig. 5.3) (Ando et al., 1990). However, this fiber diameter effect, which is only one aspect of the fiber size effect, does not convey the importance of fiber length, since it is possible to change the fiber length by, for example, a factor of two without altering either the volume fraction or the fiber diameter and the corresponding specific surface area per unit volume of mortar (as in the case of N fibers of length L initially paired and joined lengthwise to give N/2 fibers of length 2L). Such a change strongly affects the flow value at any particular fiber

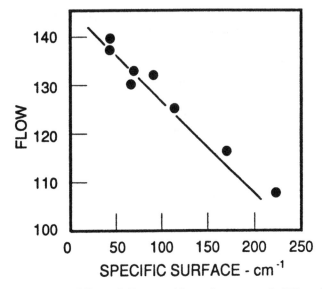

FIGURE 5.3   Effect of fiber specific surface on workability of
CFRC (Ando et al., 1990)

volume fraction with longer fibers producing more marked
mixture stiffening and reduction of flow (Fig. 5.4) (Banthia,
Moncef and Sheng, 1994).

Clearly, the combination of fiber length and diameter, as
contained in the term aspect ratio (length/diameter), still
governs the mixture-stiffening effect of a particular volume
fraction of fibers in the matrix, as described in Chapter 2. In
addition, the surface texture and possibly the flexibility of the
fibers may influence the ease with which fibers disperse and
the associated fluidity of the mixture, as in the performance of
6 mm long carbon fibers compared with 6 mm polypropylene
(Fig. 5.4), where the polypropylene fibers offer comparable
mixture flow despite a much higher aspect ratio than the
carbon fibers (1500 vs. 170) and much smaller diameter (4 μm
vs. 18 μm).

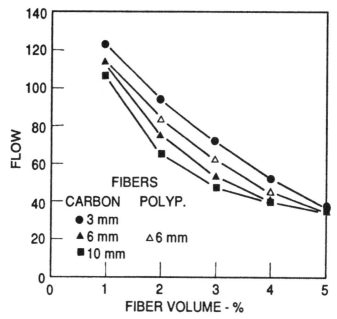

FIGURE 5.4   Effect of fiber volume fraction, length and type on workability of CFRC (Banthia, Moncef and Sheng, 1994)

5.1.4. Aramid Fiber-Reinforced Cement

Aramid fiber-reinforced cement (ARFC) employs fibers based on aromatic polyamides developed for use in tires, ropes, sails, etc.. Like carbon fibers, they are relatively expensive and difficult to disperse in a cementitious matrix using conventional mechanical mixing. However, unlike carbon fibers, they are not so easily damaged by mechanical mixing.

Silica fume in combination with a superplasticizing admixture apparently improves the dispersability of aramid fibers (Soroushian, Bayasi and Khan, 1990) in the same way as described previously for carbon fibers. Using a conventional mortar mixer, 12 μm diameter fibers in lengths of 3–13 mm, corresponding aspect ratios 250–1050, were dispersed in a superplasticized cement paste with silica fume. The mixing regime involved low speed mixing of silica fume with two thirds of the water and superplasticizer for about a minute, gradual addition of up to 2% volume of fibers with continued mixing, addition of remaining water and superplasticizer followed by cement and mixing for a further minute, stopping the mixer for 30s, and final mixing at medium speed for another minute. The mixture-stiffening effect of these rather high aspect ratio fibers in a cement-silica fume paste matrix is considerable, and causes a marked decrease in flow with increase in fiber volume fraction and length or aspect ratio (Fig. 5.5) (Soroushian, Bayasi and Khan, 1990), similar to that previously described for carbon fibers (Fig. 5.4).

5.1.5. Natural Fiber-Reinforced Cements

Numerous unprocessed natural fibers are available in many of the developing countries, and some are potentially attractive

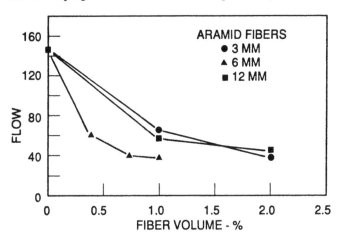

FIGURE 5.5   Effect of fiber volume fraction and aspect ratio on workability of AFRC (Soroushian, Bayasi and Khan, 1990)

as reinforcement for cement-based composites simply because of ready availability in large quantities at low cost as a continuously renewable resource (Cook, 1980). They include stem or bast fibers such as jute, flax, ramie, sunn, kenaf, urena, elephant grass, hemp, and various species of wood. However, wood fibers are more common in the developed countries after processing in a pulp mill to remove lignin leaving cellulose fibers. Such fibers can be incorporated into fiber cements prepared by the Hatschek process, as described in the next section.

Manual or mechanical mixing of unprocessed natural fibers with a cement matrix can be accomplished by premixing cement, water and additives to form a slurry, adding fine aggregate, and adding the relatively fragile fibers last (ACI Committee 544, 1996). Slump loss can be expected due to the high water absorption of 50–180% typical of many of these fibers. Alternatively, the fibers may be presaturated and added to cement, sand and admixtures with a very limited amount of water to produce a dry stiff mixture compactible only under pressure.

Most natural unprocessed fibers contain glucose which retards hardening of cement, and most are susceptible to rot as a result of bacterial or fungal action under moist conditions. An accelerating admixture may be needed to counter the set-retarding effect of the glucose, and an organic microbiocide may be needed to prevent bacterial attack of the fibers (ACI Committee 544, 1996). Most of these fibers are also prone to excessive dimensional change during wetting and drying because of their very high absorption, and are subject to deterioration by the action of alkalis in cement paste (Lewis and Mirihagalia, 1979) (Sethunarayan, Chockalingham and Ramanathan, 1989). Their long-term effectiveness depends on using fiber treatments that reduce dimensional changes during wetting or drying and using pozzolanic admixtures to lessen the severity of attack by cement alkalis.

5.1.6. Fiber Cements Prepared by the Hatschek Process

Several types of fiber, particularly those which are too fragile to withstand conventional mechanical mixing without significant damage, have been incorporated into cementitious matrices using the Hatschek process or modifications thereof summarized in 2.2.3. This process, developed originally for the production of asbestos-cement products, has been utilized and modified in various ways as the search proceeds for reinforcing fibers that can successfully replace asbestos. The matrix in this process is more dilute and fluid than in the processes previously described, with a fiber plus cement (or filler) solids content of only 5–10% when initially mixed. To retain cement and other filler fines in suspension in such dilute mixtures requires the inclusion of small high-surface-area fibers such as some wood, acrylic or polyethylene pulps. These are called filter or process fibers, and have a negligible reinforcing effect. The highly dilute mixture is picked up on a moving belt and subsequently dewatered and concentrated by pressing or vacuum suction. Many combinations of reinforcing fibers, filter fibers, matrix fillers and flocculants have been investigated (Gale, 1994).

The types of reinforcing fibers evaluated as asbestos replacements include natural cellulose-based materials derived from plants or trees (Soroushian, Shah and Marikunte, 1994) (Coutts, 1992), and a variety of synthetic materials such as polyacrylonitrile (acrylic), polyvinyl alcohol (PVA), rayon, polyester, nylon, polypropylene, polyethylene, aramid and carbon (Gale, 1994). All offer some advantages and disadvantages over the historically successful asbestos. The unique size distribution of asbestos fibers provides both small filter fibers and larger reinforcing fibers simultaneously from one source. Most alternative systems require a source of reinforcing fiber and one or more different sources of filter fiber. Also, a dispersant is usually required to help separate the fibers in the mixer, and a flocculant is subsequently needed once the fibers have been dispersed to thicken the mixture to facilitate pick-up by the moving belt. Few alternatives match the filtering ability of asbestos to retain cement particles (Gale, 1994). However, the most serious disadvantage of asbestos, and the most compelling reason for seeking alternatives, is its adverse effect on human health and the consequent bans on asbestos products pending or adopted in many countries.

5.1.6.1. Cellulose Fiber-Reinforced Cements

Cellulose fibers can be obtained from plants such as jute, coir, elephant grass, sisal and bamboo, but trees are the major source for the fibers evaluated in cellulose fiber-reinforced cements (Coutts, 1992). The fibers are obtained from various wood species by mechanical or chemical pulping, or a combination of both. In the mechanical process the fibers are separated by grinding of bulk wood often in the presence of pressurized steam. In the chemical process the fibers are separated by dissolving the natural bonding lignin with a chemical.

Wood fibers vary widely in size and structure depending on species. Softwoods as a group tend to produce longer (3–5 mm average) fibers with diameters of the order of 45 μm for springwood to 13 μm for summerwood. Hardwood fibers are shorter (1 mm average), thinner (20 μm average), and have a greater cell wall thickness and stiffness than springwood softwood fibers. The length distributions for typical softwood (pinus radiata) and hardwood (eucalyptus regnans) fibers are compared in Table 5.4 (Coutts, 1989).

As a consequence of their smaller size the number of fibers per unit mass is much larger for hardwoods, typically 7–8 times the number for softwoods (Coutts, 1992). Wastepaper, another potential source, falls between these limits depending on the constituents originally used to make the paper and the effects of recycling on fiber length. Hardwood fibers, in addition to being shorter, tend to have a higher fines content after pulping than softwood fibers, thus increasing water demand in

TABLE 5.4   Length distribution for cellulose fibers derived from
various woods (Coutts, 1989)

Length (mm)	Weighted distribution (%)		
	P. radiata	E. regnans	Wastepaper
< 0.2	2.6	2.0	6.6
0.2–0.6	4.0	12.8	21.2
0.6–1.2	9.1	74.2	40.0
1.2–2.0	18.0	9.0	15.3
2.0–3.0	28.0	1.5	11.0
3.0–4.0	25.0	–	5.2
4.0–5.0	11.0	–	1.0
> 5.0	3.0	–	–
Weighted Average (mm)	3.2	1.0	1.9

cement-based mixtures. The pulping process also influences
the purity of the resulting fibers. Chemically produced (often
termed kraft) pulps are less contaminated with lignin, which is
vulnerable to attack by cement alkalis, than mechanically produced pulps (Soroushian and Marikunte, 1990). The flexibility
of the fiber is also important for subsequent mixing, and in this
regard chemically produced fibers are much more flexible and
easier to blend with the matrix than mechanically produced
pulps (Coutts, 1992). Mechanically produced pulps are also
less suitable for autoclave curing because the high temperature and alkalinity cause them to release wood contaminants
that chemically poison the matrix creating poor interfacial
bond (Coutts, 1992). The longer chemically produced kraft
softwood pulps therefore tend to be the preferred wood fiber
from the point of view of reinforcing effectiveness and compatibility with a cement-based matrix.
   A wide range of wood pulp-based cellulose fiber-reinforced
cements is possible depending on the wood species (hard-

woods vs. softwoods), within-species differences in fiber size and cell wall thickness (springwood vs. summerwood), nature of the pulping process (chemical vs. mechanical), presence of contaminants (lignin, pulping fines etc.), water absorption capacity of the fibers, matrix constituents, dewatering pressure or suction, and nature of the curing process. All wood fibers are hygroscopic to some extent. Water absorption under moist conditions causes swelling, reduced fiber stiffness, and breakdown of the chemical bond between fibers and matrix. Consequently, composite strength is significantly reduced under wet conditions, although toughness can increase due to the increased frictional bond and pullout resistance of the fibers caused by swelling (Coutts, 1992) (Soroushian, Marikunte and Won, 1995). Therefore, either air-curing or autoclaving is preferable to normal moist curing. Certain contaminated fibers are unsuitable for autoclaving, but otherwise autoclaving seems to be the preferable method of curing. Autoclaved products are generally made with lime and silica added to the cement and offer a mature full-strength product after a few hours compared with the 2–4 weeks needed for air curing. Fiber contents up to 12% by weight are viable for these types of composite (Coutts, 1989) (Vinson and Daniel, 1990).

Fibers modified or refined to improve reinforcing effectiveness have been evaluated using chemically pulped slash pine (softwood) as the raw material (Vinson and Daniel, 1990). Refinements evaluated include fractionation to increase the proportion of stronger summerwood fibers, moderate defibrillation to break down large fiber groups into smaller ones using a laboratory beater, or very intensive processing to reduce the original raw material to virtually individual fibrils. Using a slurry dewatering Hatschek-based process, the fibers were dispersed by high speed mixing in water, cement was added with continued high speed mixing to give the desired final fiber content, the slurry was flocculated by adding an anionic polyacrylamide with gentle agitation, and then it was poured into a square mold fitted with an assembly of permeable screens, the finest being 150 μm. Excess water was removed from the 381 mm square panels by vacuum dewatering and the 13–19 mm thick panels were then pressed at 8 MPa for 3 minutes to remove further water prior to curing at 100% relative humidity for 7 days and subsequently at 50% relative humidity until testing. Fines retention varied considerably with 1% to 18% loss of cement, so the fiber type and the nature of the refinement is important from a processing viewpoint. Fiber type was less important than fiber content in determining final water-cement ratio and density. Generally, water-cement ratio increased and densities decreased correspondingly with increasing fiber content (Table 5.5).

5.1.6.2. Other Synthetic Fiber-Reinforced Cements

Synthetic fiber-reinforced cements produced commercially using the Hatschek process most commonly employ polyvinyl alcohol (PVA), polyacrylonitrile (acrylic), or to a lesser extent polypropylene as the reinforcing fiber (Gale, 1994). One or more types of filter or process fiber are also included in what are often proprietary mixtures kept secret by the manufacturers. Extrusion technology has also been investigated as an alternative to the Hatschek process. This involves forming thin-section products by forcing a highly viscous dough-like mixture through a shaping die under high pressure (Shao, Marikunte and Shah, 1995).

The primary objective is to find a fiber that can effectively substitute for asbestos in the production of thin-section fiber cement products by the Hatschek process. Studinka (1989) discusses the search for an asbestos substitute, and concludes that, although none could be found that matched asbestos in all respects, PVA fiber reinforcement produced the best combination of properties. The PVA fibers in fiber cements produced by Akers, Studinka et al. (1989) using cellulose pulp as the process fiber were found to be unaffected by the cement alkalis or the Hatschek process. However, the cellulose-based

TABLE 5.5 Mixture characteristics for experimental refined cellulose fiber-reinforced cements (Vinson and Daniel, 1990)

Description of slash pine softwood fiber	Fiber % by wt.	Cement loss — %	w/c	Density (dry) — kg/m³
Natural[a], beater-refined	4	1.0–1.5	0.33	1894
	8	1.0–1.5	0.34	1733
	12	1.0–1.5	0.43	1573
Summerwood enriched[b], unrefined	4	17–18.5	0.34	1910
	8	17–18.5	0.39	1749
	12	17–18.5	0.57	1541
Summerwood enriched[b], beater-refined	4	1.8–2.6	0.33	1974
	8	1.8–2.6	0.36	1798
	12	1.8–2.6	0.40	1669
Natural[a], intensively defibrillated	4	Negligible	0.30	1974
	8	Negligible	0.35	1733
	12	Negligible	0.42	1621

[a] 55% summerwood fibers.
[b] Fractionated in laboratory to achieve 86% summerwood fibers.

process fibers are prone to natural weathering (Akers and Studinka, 1989). To address the potentially inferior weathering characteristics of cellulose, blends of polyethylene pulp or polyoxymethylene pulp were used with softwood kraft pulp as the process fiber in evaluating acrylic fiber-reinforced cements, in the hope that reducing the amount of cellulose-based process fibers would lead to improved durability (Table 5.6) (Daniel and Anderson, 1986). The mixing sequence with a high-speed shear mixer consisted initially of combining a dispersion of process fibers in water (2%, 3% and 4% by weight of water for polyethylene, polyoxymethylene, and kraft pulp respectively) with cement at a water-solids ratio of about 10:1 by volume. Then the 6 mm long acrylic reinforcing fibers (1 to 3% by weight of total ingredients) were mixed into the initially fluid matrix to obtain a slurry of adequate consistency. Finally, an anionic flocculant was mixed into the slurry by hand, prior to fabrication of sheets using vacuum-dewatering (127 mm of mercury) and pressing (7 MPa) stages typical of the Hatschek process to achieve a final water-cement ratio of about 0.30.

Certain forms of polyethylene pulp, in addition to imparting the filter characteristics of a process fiber needed to retain cement in suspension, are reported to also have significant reinforcing effectiveness (Gale, Shah and Balaguru, 1990).

TABLE 5.6   Combinations of acrylic reinforcing fibers and process fibers (Daniel and Anderson, 1986)

Fiber type	Amount-percent by weight of fibers									
Acrylic reinforcement	1.0	2.0	3.0	1.0	2.0	3.0	2.0	2.0	2.0	2.0
Process fiber										
Polyethylene pulp				1.5	1.5	1.5			1.5	1.5
Polyoxymethylene process	1.5	1.5	1.5				1.5	1.5		
Kraft cellulose pulp	1.5	1.5	1.5	1.5	1.5	1.5	2.0	2.5	2.0	2.5

These pulps consist of very fine short monofilaments of irregular shape and surface texture sufficient to promote mechanical bonding over and above the rather low adhesive bond associated with smooth monofilaments. They are prepared by a special spinning and refining process which produces a high level of molecular orientation that approximately triples the material strength compared with unoriented equivalents. These fibers of length about 1 mm and wide ranging diameter (1–20 μm) are claimed capable of acting as both reinforcing fibers and filter or process fibers for keeping cement in stable suspension. They may also be combined with other reinforcing fibers such as PVA (polyvinyl alcohol) or PAN (acrylic) to produce a cellulose-free composite with consequent potential for improved durability (Gale, Shah and Balaguru, 1990).

5.2. FIBER-REINFORCED CONCRETES

Unlike fiber-reinforced cements, the matrix in these composites contains a significant volume fraction of coarse aggregate larger than 5 mm, and consequently a much smaller paste volume fraction available to accommodate the fibers. Therefore, the maximum fiber amount possible without excessive mixture stiffening and loss of workability is correspondingly smaller, often less than 1%, and usually not more than 2% by volume of mixture. Rigorous mechanical mixing is usually needed to blend the fibers with coarse and fine aggregates, cement, and any filler materials present. Sometimes, fly ash, slag or silica fume are used to increase the paste volume fraction and facilitate accommodation of fibers. Only the more robust types of fiber are capable of withstanding the bending, impact and abrasive effects inherent in rigorous mechanical mixing. Steel fibers are most commonly used, but fibrillated polypropylene, and monofilament forms of polypropylene, polyethylene, polyester, polyvinyl alcohol and polyacryloni-

trile can also be incorporated into concrete by mechanical
mixing, or in some cases by dry-process shotcreting. The
length and crossectional size of fibers used in concrete are
generally larger than the corresponding fiber parameters used
in fiber cements, consistent with evidence suggesting that fiber
length should exceed the maximum aggregate size in the
matrix as discussed in 3.2.6.

5.2.1. Mechanically Mixed Steel Fiber-Reinforced Concrete

The fibers for steel fiber-reinforced concrete (SFRC) are manufactured in many mostly monofilament forms, often with
modifications intended to increase their resistance to pullout
from the matrix, as described in 3.2, that may influence their
miscibility with concrete. The concrete matrix for mechanically mixed SFRC needs to be proportioned so that satisfactory workability is achieved with a minimum amount of water
while keeping the maximum aggregate size less than the fiber
length. Smooth rounded sand with a low fineness modulus and
a predominance of particles passing the 600 μm and 300 μm
sieves is preferable to a coarse angular manufactured sand
(Schraeder 1988). Coarse aggregate maximum size should not
usually exceed 25 mm, although mixtures with 38 mm aggregate have been made satisfactorily (Tatro, 1987), again using
sand with a high proportion passing 600 μm and 300 μm
sieves. Reducing the coarse aggregate volume fraction, as in
concrete proportioned for pumping, allows the fine aggregate
and mortar volume fractions to increase when cement and
water are kept constant. Alternatively, the volume of the paste
can be increased without increasing the sand by use of fly ash,
slag or silica fume to facilitate accommodation of the fibers
and improve workability at any particular fiber content as
shown previously in 2.1.2 (Fig. 2.3).

To minimize water demand for any specified level of workability, addition of a supplementary cementing material like fly

ash is preferable to an excess of cement or sand or both, because of the more rounded particle shape of fly ash. Water added to compensate for the mixture-stiffening effect of the fibers should be limited to the amount needed to reach the maximum water-cement ratio permissible for the applicable exposure condition, typically 0.40 to 0.55. Use of water-reducing or superplasticizing admixtures helps to meet both workability and water-cement ratio requirements. Guidelines for water-cement ratio, cement content and fine/total aggregate percentage have been developed (ACI Committee 544, 1993) (Table 5.7).

Any approach to proportioning SFRC for workability must recognize that the result depends on the method of assessing workability and that the slump test which is the dominant method of measuring workability for concrete without fibers may not be satisfactory for many fiber-reinforced concretes. The reason is that the slump test tends to exaggerate the mixture-stiffening effect of fibers because it assesses primarily the stability or cohesion of the mixture under static conditions, while in practice it is mixture behaviour under the dynamic

TABLE 5.7 Recommended range of mixture proportions for SFRC (ACI Committee 544, 1993)

Mixture characteristic	Coarse aggregate maximum size		
	10 mm	20 mm	38 mm
Water/cement	0.35–0.45	0.35–0.50	0.35–0.55
Cement-kg/m^3	360–600	300–540	280–420
Fine/total agg. — %	45–60	45–55	40–55
Entrained air — %	4–8	4–6	4–5
Smooth fibre — %[a]	0.8–2.0	0.6–1.6	0.4–1.4
Deformed fibre — %[a]	0.4–1.0	0.3–0.8	0.2–0.7

[a] Percent by volume of concrete. 1%, 78.3 kg/m^3.

vibratory conditions of placement commonly used in construction practice that is most relevant. This depends primarily on the mobility and compactability of the mixture (Johnston, 1984–2). Consequently, stiffer SFRC mixtures tend to exhibit unacceptably low slump when compared to mixtures without fibers that have the same workability when judged by any test performed using vibration.

One form of such a test, ASTM Standard C 995 (ASTM, 1994), was developed to assess the mobility of SFRC in terms of the time of flow of a sample of mixture through an inverted cone (the slump cone) under internal vibration (Fig. 5.6, left). It shows that stiffer SFRC mixtures exhibit less slump than plain concretes of the same workability under vibration when judged according to inverted cone time (Fig. 5.6, right) (Johnston, 1994–1) (ACI Committee 544, 1988–1).

Another test that measures primarily the mobility and secondarily the compactability of a mixture under vibratory conditions representative of normal placement practice is the V-B test long standardized in Europe. The observation that the results of the V-B and inverted cone tests correlate strongly and linearly with each other, and that the relationship passes through the origin, suggests that both tests assess the same rheological characteristics, primarily mobility or flow under vibration and secondarily compactability under vibration (Fig. 5.7, left) (Johnston, 1994–1).

Correlations of either V-B or inverted cone time with slump (Fig. 5.7, right) are less meaningful since the slump test measures a different aspect of rheological behaviour, primarily stability or cohesion under static conditions, at least for stiffer mixtures with slumps less than 50 mm where the precision of slump measurements is poor (Johnston, 1984–2). However, for more fluid SFRC mixtures, typically obtainable using superplasticizing admixtures, both the inverted cone and V-B times often become too short (less than 3s) to be determinable with reasonable precision. On the other hand, slump increases to more than 50 mm where it is readily determinable with better

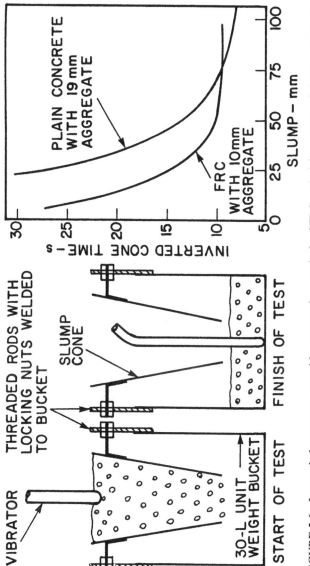

FIGURE 5.6 Inverted slump-cone test with comparative trends for SFRC and plain concrete (Johnston, 1994–1) (ACI Committee 544, 1988–1)

116

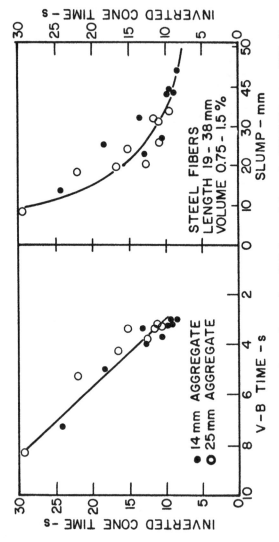

FIGURE 5.7 Relationships between inverted cone time, V-B time and slump for SFRC (Johnston, 1994–1)

precision. Also, since there is more change of sample shape at higher slump values the test result may reflect mobility more than stability, albeit under static rather than vibratory conditions. For such mixtures the slump test is probably the only practical alternative.

For any particular matrix, the mixture-stiffening and workability-reducing effect of fibers in mechanically mixed concretes primarily reflects the influence of fiber content and aspect ratio, as discussed in 2.1.1. For steel fibers in concrete, workability decreases with increase in either fiber content or aspect ratio (Fig. 5.8) (Johnston, 1994–1), just as was the case for fiber-reinforced cements (Fig. 5.4 and 5.5).

Changing the matrix by reducing the coarse aggregate volume fraction and correspondingly increasing the fine aggregate, while keeping the volume fraction and water-cement ratio of the paste constant in accordance with North American practice (ACI Committee 211, 1988) improves workability, particularly at higher fiber contents approaching the maximum possible for reasonable workability (Fig. 5.9) (Johnston, 1994–2). This is consistent with the practice of reducing the coarse aggregate volume fraction by up to 10% in plain concrete to facilitate pumping. The maximum aggregate size is also important as discussed in 2.1.3. However, although reducing it helps to accommodate fibers, other considerations unrelated to fibers usually dictate the final choice.

Recommended maximum fiber contents possible without encountering workability and fiber balling problems therefore reflect the effects of fiber aspect ratio (Fig. 5.8), coarse aggregate volume fraction (Fig. 5.9) and coarse aggregate maximum size (Fig. 2.6). The pattern in the recommendations of one major fiber manufacturer is typical of how these variables influence fiber content, and adjustments for pumped concrete are also included (Table 5.8). For other steel fiber types, the pattern should be similar, but with modifications to the actual numbers to reflect differences in fiber profile, surface texture and nature of end anchorage.

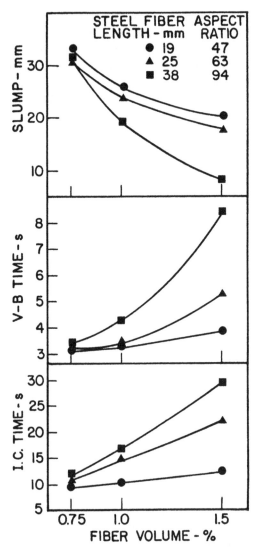

FIGURE 5.8   Effect of fiber amount and aspect ratio on the workability of SFRC (Johnston, 1994–1)

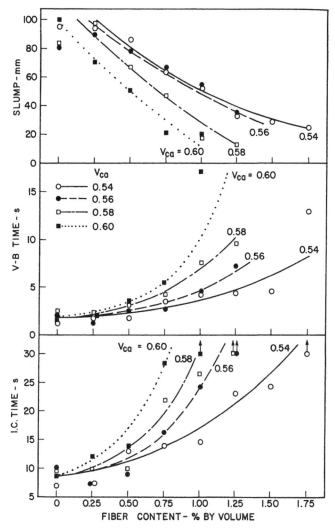

FIGURE 5.9   Effect of coarse aggregate volume fraction on the workability of SFRC (Johnston, 1994–2)

TABLE 5.8   Maximum recommended[a] steel fibre contents[b] for SFRC-kg/m^3

Max. coarse aggregate size — mm	Steel fibre aspect ratio					
	60		75		100	
	Normal	Pumped	Normal	Pumped	Normal	Pumped
4	160	120	125	95	95	70
8	125	95	100	75	75	55
16	85	65	70	55	55	40
32	50	40	40	30	30	25

[a] N.V. Bekaert S.A.
[b] 1% by volume of concrete = 78.3 kg/m^3.

Balling or clumping of fibers is a problem that can occur both in the packaged fibers before mixing, termed dry balling, or in the freshly mixed SFRC after addition of fibers, termed wet balling. Prior to entering the mixer the fibers must be free of balls that result from tangling during densification for packaging, or simply from inherent fiber characteristics such as high aspect ratio, crimping, surface roughness or end anchorage that may promote entanglement. This problem of dry balling may be addressed by passing the fibers through a 50–100 mm mesh screen, preferably connected to a vibrator. One manufacturer provides a vibro-pneumatic system consisting of a vibrating screen on top of a hopper, a blower, and a discharge pipe. The vibrating mesh separates any fiber balls, and the air blower conveys the separated fibers through the discharge pipe to the mixer or aggregate conveyor belt.

Another approach for reducing the potential for dry balling prior to mixing is bundling of the fibers with a water-soluble glue (Fig. 3.13) which decreases the effective aspect ratio of the bundled units and lessens their tendency to ball or tangle. However, for such fiber bundles the mixing time must be at least sufficient to ensure complete separation of the bundles into individual monofilaments and yet not long enough to introduce the possibility of wet balling of the individual monofilaments in the mixture due to overmixing.

Wet balling where individual monofilaments that are initially separate start to ball in the mixture can happen whether the fibers are initially bundled or simply batched in monofilament form if the recommended maxima for fiber content and aspect ratio, typified by Table 5.8, are exceeded. Other causes of wet balling are adding fibers too quickly to a matrix of too low workability initially, selecting a matrix with a coarse aggregate volume fraction that is too high, or using a mixer with worn blades or a design inappropriate for use with fibers as discussed later in this section.

The packaging and batching of fibers can be made easier by special equipment. Some manufacturers provide convenient and mobile conveyors for transporting fibers packaged in small (up to 40 kg) bags to the mixer or aggregate conveyor belt. Others provide an integrated system developed for large volume production in a mixing plant that handles packages of up to 1000 kg of fibers, provides hopper storage up to 2000 kg, vibratory separation of fibers, automatically controlled weigh-batching, and conveyance of the weighed fibers to the mixer or aggregate conveyor belt (Fig. 5.10) (Nielsen, 1994).

Another way of dispensing fibers into the mixer that avoids the need for fiber packaging, handling of packages, and separation of the fibers after opening of packages, is to eliminate the need for precut manufactured fibers by cutting fibers from wire coils fed to the mixer. An example of this technique is a system that feeds four coils of wire through an automated high speed cutter. It processes approximately 9000 m of wire per minute to feed 30–40 kg of steel fibers per minute directly into a stationary central or truck mixer (Skarendahl, 1994).

The preferred method of adding fibers to concrete ingredients is at the aggregate conveyor belt which blends them into the aggregates prior to mixing with cement and water (Fig. 5.11,

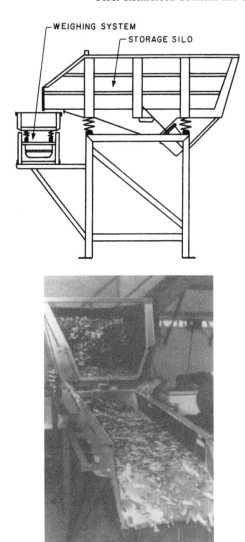

FIGURE 5.10   Integrated system for separating, weighing and
batching steel fibers consisting of vibrating feeder
silo and vibrating weigh hopper conveyor with variable controlled discharge rate (Courtesy of Skako
Ltd.)

FIGURE 5.11   Fiber batching and discharge on to the aggregate
conveyor belt prior to mixing (upper) and fiber entry
directly to the mixer by conveyor belt or blower
(lower) (Courtesy of Bekaert Corporation)

upper). Simultaneous feeding of aggregates, fibers, cement and
water into the mixer is also effective. When preblending of
fibers with mixture ingredients is not possible, they can be
added to premixed concrete (Fig. 5.11, lower) provided it has
appropriate workability, which according to various fiber manufacturers means a slump of at least 120 mm and as much as
200 mm for higher fiber amounts and aspect ratios. The initial
slump of the premixed concrete before addition of fibers
should be achieved using chemical admixtures rather than
excess water and should be 50–75 mm more than the desired
final slump (ACI Committee 544, 1993). However, in view of
the previous discussion about the validity of the slump test for
low-workability mixtures, this recommendation is probably

appropriate only for final slumps greater than about 50 mm where V-B and inverted cone times are too small to be accurately determined.

Successfully achieving uniform fiber distribution throughout the concrete matrix also depends on the type of mixer and the mixing regime. Mixers unsuitable for low-workability concrete in general are not likely to work well for SFRC. Constant flow drum-type or screw-type mixers where the materials move axially through the mixer as mixing progresses may also be unsuitable. Pan mixers where the blades rotate in the same direction as the pan are less suitable than mixers where the blades and pan rotate in opposite directions. Addition of fibers in any manner that fails to spread them throughout the batch may lead to poor fiber distribution. For example, adding fibers to the rear end of a truck mixer, rather than using a conveyor to blow or spread them along the length of the batch, tends to prolong the mixing time required for uniform distribution, and may lead to wet balling in zones of high fiber content. Regardless of the mixer type, the fiber addition and mixing regime must ensure uniform fiber distribution is achieved without overmixing and the consequent possibility of wet balling.

5.2.2. Mechanically Mixed Synthetic Fiber-Reinforced Concretes

The fibers for these concretes are typically polypropylene, polyester, nylon, polyacrylonitrile (acrylic) or polyolefin in the form of separate monofilaments, bundled monofilaments, or in the case of polypropylene more commonly in the form of multifilament fibrillated strands. The concrete matrix used with these fibers needs to be proportioned with a larger than normal mortar fraction. This means reducing the coarse aggregate volume fraction and increasing the fine aggregate or adding fly ash, silica fume or slag as discussed previously for steel fibers.

5.2.2.1. Polypropylene Fibers

Fibrillated polypropylene strand is widely available in lengths of 19 to 63 mm and in flat or twisted configurations (Fig. 3.13). Each fibrillated strand is intended to separate during mechanical mixing into ministrands comprising a few longitudinal filaments crosslinked by transverse filaments to form an open lattice that is penetrated by the matrix mortar, thus creating a mechanical bond. This mechanical bond is desirable since polypropylene is hydrophobic and therefore not easily wetted by cement paste to naturally develop an adhesive bond. However, proprietary surface treatments are used by some manufacturers to try and improve adhesive bond.

Monofilament polypropylene is available in bundles that disperse in concrete during mixing. They may be crimped to improve bond with cement or coated to improve wettability and dispersability. The mixture-stiffening and workability-reducing effect of polypropylene fibers, whether monofilament or fibrillated, is very marked due to their large surface area. It depends on fiber amount and fiber/strand length (Fig. 5.12) (Johnston, 1994–1 and 1996), just as for steel fibers (Fig. 5.8). Both forms of polypropylene are used at volume fractions as low as 0.1%, mainly to control plastic shrinkage cracking with little effect on hardened concrete properties. The fibrillated forms have been used in larger amounts up to 0.7% by volume where hardened concrete properties can be significantly improved.

To achieve proper strand separation of the fibrillated forms of polypropylene with adequate length for bond development and uniform fiber distribution in the mixture, two manufacturers recommend strand lengths related to aggregate maximum size. However, the recommendations differ because of differences in the raw material and the method of forming the strands. In one case the fiber length is 1.5–2.0 times the coarse aggregate maximum size, while in the other it is about three times the aggregate size (Table 5.9). The manufacturer's

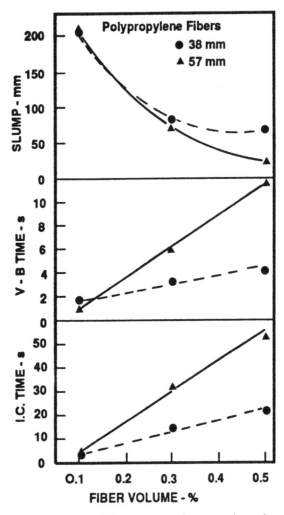

FIGURE 5.12   Effect of fiber amount and aspect ratio on the workability of concrete with fibrillated polypropylene
strand (Johnston 1994–1 and 1996)

TABLE 5.9   Manufacturers recommended strand lengths for fibrillated polypropylene

Aggregate[a] size — mm	Strand[a] length — mm	Aggregate[b] size — mm	Strand[b] length — mm
5	6–13	6	19
10–16	19	13	38
16–25	38	19	57
19–38	51	25+	63

[a] Fibermesh flat strand.
[b] Forta twisted strand.

reported experience in the latter case shows that using long strand with small aggregate, for example 63 mm strand in a mortar matrix, leads to mixing and distribution problems probably caused by the small aggregate particles being unable to effectively separate the filaments in the long strands. An investigation using tilting drum, spiral flow and pan-type mixers, with examination of strands recovered following wash-out after mixing, showed that the nature of the strand, the type of mixer, and the mixture slump affect the percentage of strands judged to have open properly. The pan mixer was most effective at both zero and 100 mm slumps with mixing times up to 100s, and separation of strands was more complete at 100 mm slump than at zero slump for all mixtures (Fig. 5.13) (Nanni and Meamarian, 1991). The proportion of strands sheared (reduced to smaller than 30% of original size) during mixing was also assessed and found to be substantial after 60s mixing time for the pan and spiral flow mixers. The tilting drum mixer was less severe with respect to shearing the strands. Clearly, an effective mixing system must separate the strands reasonably completely without excessively shearing them and reducing aspect ratio as a consequence.

An integrated system is available to cut strand supplied in coils, and separate, weigh and batch the cut strand into the

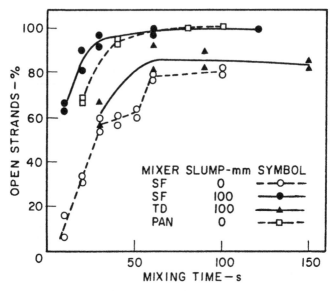

FIGURE 5.13   Effect of mixer type and mixture slump on separation of fibrillated polypropylene strands (38 mm
long, 0.2% by volume) during mixing (Nanni and
Meamarian, 1991)

concrete (Fig. 5.14). Otherwise, strands or bundles can be
added with other ingredients as they enter the mixer, or to premixed concrete provided they are uniformly distributed
throughout the batch or load. In the latter case, the initial concrete workability before adding fibers should be high, at least
100 mm slump for 0.1% fibers and progressively more for
greater fiber amounts, preferably by use of chemical admixtures rather than excess water. However, the manufacturers
recommend that when a superplasticizer is used the fibers
should be blended with the matrix before the superplasticizer
is added.

Due to the highly cohesive nature of mixtures with
polypropylene fibers, their workability appears very low when
judged on the basis of slump, and is accordingly more realisti-

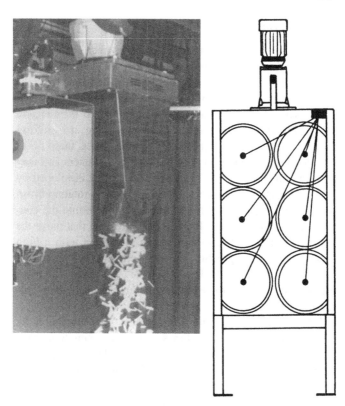

FIGURE 5.14   Integrated system for multiple fiber rolls, fiber
cutting, weighing and batching synthetic fibers
(Courtesy of Skako Ltd.)

cally evaluated by V-B time or inverted cone time for the
reasons discussed in the previous section. Unfortunately, the
inverted cone test can sometimes be rendered invalid by long
fibers wrapping around the vibrator, or the mixture failing to
exit the cone after unlimited vibration when it is very cohesive, leaving a hole in the test sample with the remainder stuck
to the sides of the cone.

5.2.2.2. Polyacrylonitrile (acrylic) Fibers

Some acrylic fibers are produced both for fiber cements prepared by the Hatschek process, as discussed previously in 5.1.6.2, and in larger versions (52–104 μm diameter) for addition to conventionally mixed concrete. Sometimes they can be added without any special means of separation if the concrete mixer effectively induces separation during the mixing process. However, like other monofilament types, these fibers tend to clump together in their packaging containers and may require a fiber separating machine, such as that developed by one manufacturer which employs a high speed rotating brush that loosens the bundled fibers and injects them into the concrete mixer (Wörner and Techen, 1994). Mixers that move the matrix on the counterflow principle or mixers with a high speed, high shearing action are more effective for separating the fibers, especially the longer ones (12–25 mm), while less efficient mixers need to be used in conjunction with fiber separation equipment.

Fiber dispersion in the matrix must be achieved quite quickly within short mixing times, as longer mixing times may cause wet balling. Larger low-aspect-ratio fibers are easiest to disperse, as for other fiber types. Typical fibers for use in concrete are 6 or 12 mm long and 0.1 mm diameter, and water-reducing or superplasticizing admixtures are recommended to reduce the mixture-stiffening effect of the fibres and avoid excessively high water-cement ratios. Fiber contents in the range 5–17 kg/m^3 (0.5–1.5% by volume) have been used in practice in Europe (Wörner and Techen, 1994).

Various mixing regimes have been evaluated for concretes with different maximum aggregate sizes using 6x0.1 mm fibers. They are identified in Fig. 5.15 (Wörner and Techen, 1994), as follows:

(a) mixing the concrete followed by addition of fibers

(b) mixing fine aggregate, cement and water, addition of fibers, and finally addition of coarse aggregate.

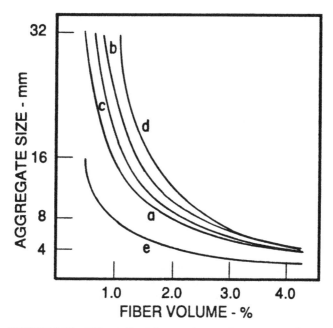

FIGURE 5.15   Effect of mixing regime and aggregate size on
maximum volume percentage of acrylic fibers
(Wörner and Techen, 1994)

(c) as for (a) but using a fiber separator

(d) as for (b) but using a fiber separator

(e) dry-mixing of cement and fibers followed by normal
wet mixing with aggregates and water.

The results in Fig. 5.15 indicate the limiting fiber volume
fractions for 6 × 0.1 mm fibers of aspect ratio 60 according to
aggregate maximum size. They show that the largest fiber contents are achieved with regimes (b) and (d) which involve premixing the mortar fraction, adding fibers, and finally adding
coarse aggregate. Using these regimes with, for example,
16 mm aggregate limits the fiber amount to about 1.8% by

volume or about 20 kg/m³, a maximum volume percentage not very different for similar concrete with steel fibers of the same aspect ratio.

5.2.2.3. Polyolefin Fibers

Fibers categorized as polyolefins include polypropylene, polyethylene and higher molecular weight hydrocarbon-based polymers. High-density polyethylene (specific gravity 0.96) fibers produced in 40 × 0.9 mm monofilaments surface-roughened with wart-like deformations have been used in Japan (Kobayashi and Cho, 1981) in mechanically mixed concrete with 15 mm aggregate in amounts of 2–4% by volume of concrete. The apparently large 4% limit suggested as the practical maximum for mixing is consistent with the relatively low fiber aspect ratio of 44.

Recently, a form of fibers described only as polyolefin with specific gravity 0.91, close to that of polypropylene, has become available in North America. They are smooth monofilaments with a proprietary surface treatment to improve bond, and are grouped in parallel and wrapped with tape to form cylindrical bundles about 55 mm in diameter (Fig. 5.16). The wrapping tape is formulated to be water-dispersible, thus providing a timed release that allows the bundle to separate into individual fibers as mixing progresses. The bundles are reported to distribute throughout the batch during the first 3–4 minutes of mixing (Morgan and Rich, 1996). Then the tape releases allowing them to separate into monofilaments that disperse in the mixture during the remaining 3–4 minutes of mixing. Using 25 × 0.38 mm fibers of aspect ratio 66, fiber amounts up to 2% by volume (18.2 kg/m³) were successfully mixed in a transit mixer for subsequent placement by the wet-mix shotcreting process (Morgan and Rich, 1996). Using higher aspect ratio 50 × 0.63 mm fibers of aspect ratio 80 for transit-mixed, cast-in-place field trials, fiber amounts of 12–15 kg/m³, 1.25–1.65% by volume, could be satisfactorily incorporated in concrete (Ramakrishnan and Kakodkar, 1995).

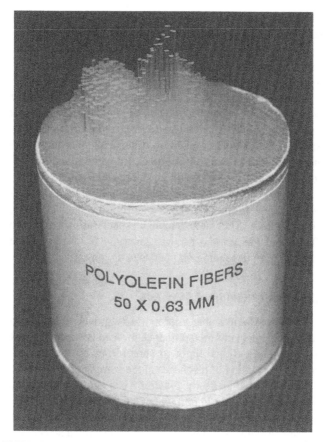

FIGURE 5.16   Paper-wrapped polyolefin fiber bundles (Courtesy of 3M Company)

5.2.2.4. Other Polymeric Fibers

Polyester is a term encompassing a variety of material groups ranging from the textile fibers developed in the 1950's to thermoplastic types marketed for use in concrete primarily to control plastic shrinkage cracking. Some polyesters are reported as subject to deterioration in cement due to hydrolysis and dissolution of the basic ingredient, polyethylene tereph-thalate, in the presence of cement alkalis (Wang, Backer and Li, 1987) (Jelidi et al., 1991). There is little consensus on the long-term compatability of polyester fibers with cement-based matrices.

Nylon is a generic name for a group of amide-based poly-mers. Two types, nylon 6 and nylon 66, are currently marketed for use in concrete primarily to control plastic shrinkage cracking. Nylon is hydrophilic to water with an absorption of about 4%, and is resistant to attack by cement alkalis (ACI Committee 544, 1996) (Khajuria, Bohra and Balaguru, 1991).

Both nylon and polyester fibers are supplied in bundled monofilament forms that disperse during mechanical mixing with concrete. The monofilament aspect ratio is typically 60–80, and the amount used is generally 0.1% by volume, 1.1–1.4 kg/m^3 depending on fiber specific gravity.

5.2.3. Fiber-Reinforced Shotcrete

The principles applicable to proportioning, batching and pro-ducing mechanically mixed fiber-reinforced concrete for cast-in-place applications apply to shotcreting by the wet-mix process. Prevention of blockages in the pumping system due to dry or wet fiber balls by placing a screen over the pump hopper is particularly important (ACI Committee 506, 1984).

Likewise, when using the dry-mix process, blockages in the spraying system should be prevented by placing a screen over the receiving hopper to intercept any balls formed in the dry mixture prepared by first mixing fibers with fine aggregate fol-lowed by cement and other ingredients. It has also been found that a good electrical ground to the spray gun and nozzle reduces fiber clumping and plugging at the nozzle when using steel fibers. Screw-type mixers have been found satisfactory for mixing the dry ingredients prior to discharge into the shot-crete hopper (ACI Committee 506, 1984). However, fibers other than steel have not been found very suitable for the dry-mix process.

Generally, fibers suitable for shotcreting have a shorter length and lower aspect ratio than those used in mechanically mixed fiber-reinforced concrete for cast-in-place application. These characteristics facilitate production of ball-free material and ease of passage through pipes or hoses. For either dry or wet-mix processes the pipe or hose diameter should be at least 1.5 times the fiber length and not less than 50 mm.

5.2.3.1. Steel Fiber Shotcrete

Fiber contents are in the range 40–80 kg/m^3, 0.5–1.0% by volume. Silica fume is often used to reduce fiber rebound, particularly in the dry-mix process where rebound tends to be greater (20–50%) than in the wet-mix process 5–10% (Morgan, 1991). In the dry-mix process there is evidence of significant preferential 2-dimensional fiber alignment in the plane of the sprayed section (Armelin and Helene, 1995), which is advantageous for in-plane mechanical properties, but may be offset to some extent by greater fiber rebound. Even in the wet-mix process, some degree of in-plane fiber alignment seems inevitable as a result of spraying, and mechanical properties are likely less adversely affected by fiber loss due to rebound. Typical compositions for dry and wet process mixtures are given in Table 5.10 (Morgan, 1991) (Morgan, et al. 1992).

A variation of the wet-mix process that permits the use of fiber aspect ratios greater than the 50–70 range in Table 5.10 involves cutting of long fibers from two coils of wire fed into the spray gun and injecting them into the pumped premixed concrete matrix at the gun (Fig. 5.17) (Skarendahl, 1992). It eliminates the need for precutting, packaging and separation of fibers prior to mixing, and of course the higher aspect ratio improves their reinforcing effectiveness.

5.2.3.2. Polypropylene Fiber Shotcrete

Both the monofilament and fibrillated forms of polypropylene have been utilized to a limited extent in both dry and wet-mix

TABLE 5.10 Typical fiber shotcrete mixture characteristics

Constituent material	Steel fibers[a]		Polypropylene fibers[b]		Polyolefin fibers[c]
	Dry-mix	Wet-mix	Wet-mix	Wet-mix	Wet-mix
Cement — kg/m³	400	420	402	157	400
Fly ash — kg/m³	–	–	–	236	–
Silica fume — kg/m³	50	40	–	–	48
10 mm. aggregate — kg/m³	500	480	430	403	480
Concrete sand — kg/m³	1170	1120	1285	1230	1110
Fibres — kg/m³	60	60	4	6	9–12
Water reducer — l/m³	–	2	2	–	1.8
Superplasticizer — l/m³	–	6	–	7	1
Water — kg/m³	170	180	190	162	190
Air-entraining admixture	No	Yes	Yes	Yes	Yes

[a] Aspect ratio 50–70.
[b] 38 mm fibrillated strand.
[c] Bundled monofilaments, aspect ratio 66.

FIGURE 5.17   Spray gun with dual wire feed for long fiber shotcrete (Courtesy of Å. Skarendahl)

processes at low volume fractions, typically 0.1% volume
(0.9 kg/m³). However, most recent work has involved wet-mix
application with higher volume fractions of 0.4–0.7%
(4–6 kg/m³) where improvements in mechanical properties are
likely to be more significant. Increasing the paste volume fraction by using fly ash replacement of cement up to 60% of total
cementitious material has been shown effective for raising the
maximum possible fiber content from 4 kg/m³ to 6 kg/m³
(Table 5.10) (Morgan et al., 1992).

Typically, wet-mix polypropylene fiber shotcretes
(Table 5.10) are made with 10 mm aggregate, sand, cement,
water-reducing or superplasticizing and air-entraining admixtures, and 4–6 kg/m³ of 38 mm fibrillated strand (Morgan et
al., 1992) (Morgan et al., 1989).

5.2.3.3. Polyolefin Fiber Shotcrete

The bundled form of monofilament polyolefin fibers described previously has been utilized in the evaluation of wet-mix shotcrete using 25 × 0.38 mm fibers of aspect ratio 66 (Morgan and Rich, 1996). Although the tests included fiber amounts up to 2% by volume (18 kg/m^3), amounts of 1.0–1.5% (9–14 kg/m^3) are recommended as optimum from the point of view of pumpability and improvement in mechanical properties. The mixtures where prepared at 10 ± 1% air before pumping to achieve an in-place air content after shotcreting of 4 ± 1%.

5.2.4. Slurry Infiltrated Fiber-Reinforced Concrete

The process of depositing fibers in a mold and infiltrating them with a cement-based matrix is not readily suited for any kind of mechanized large-volume manufacturing process. Generally, the fibers are manually sprinkled on the forming surface or into a mold in their natural packing array, and are levelled using a hand rake followed by a metal screed positioned on top of the formwork. In a slab or overlay application they can be manually rolled using a conventional lawn roller or equivalent to orient the top fibers parallel to the wearing surface, thus minimizing fiber protrusions (Lankard and Newell, 1984). The final fiber content depends on fiber aspect ratio and other characteristics which influence the packing density, but is typically 8–12% by volume for steel fibers, much higher than the 1.5–2.0% maximum achievable with mechanically mixed fiber-reinforced concrete or shotcrete.

The matrix is generally a cement paste or mortar with fly ash or silica fume, superplasticizing admixture, and possibly latex emulsion (Naaman, 1992) (Lankard, 1986). The maximum sand particle size in the mortar must be small enough to avoid clogging in the fiber bed that inhibits complete infiltration of the matrix slurry (Lankard, 1986). Water-cement or water-cementitious materials ratios are usually 0.20–0.45, and sand-cementitious material ratios range from zero to 1.0 (Schneider, 1992). Fabrication involves gravity pouring of the slurry taking care to expel air by maintaining a continuous advancing front of slurry (Fig. 2.11). In an overlay application the slurry was poured through a grate with vibrators attached to it that was positioned on top of the formwork at the level of the finished surface (Lankard and Newell, 1984).

CHAPTER 6

Properties and Applications — Fiber-Reinforced Cements

Modern fiber-reinforced cements are used mainly for thin sheet-like products that are in many cases intended for applications historically serviced by asbestos-cement. These include flat and corrugated sheets, siding or cladding panels for buildings, shingles, shakes and slates for roofing, and various kinds of pipe for use in non-pressurized water and sewer drainage or for transmission of fluids under pressure.

The property requirements for various categories of application are identified in performance specifications published in different countries. The specifications discussed in this chapter reflect mainly North American practice as published in ASTM standards, recognizing that, while standards in other parts of the world may differ in detail, they are often similar in principle with respect to the properties that must be evaluated to establish material conformance.

6.1. SPECIFICATIONS FOR ASBESTOS-CEMENT

ASTM specifications for asbestos-cement products (ASTM, 1995), first published in 1949, are relevant mainly from a historical perspective. They include ASTM C 220 (flat sheets), C 221 (corrugated sheets), C 222 (roofing shingles), and C 223 (siding). All four specifications purport to specify minimum flexural strength requirements, but actually embody flexural load requirements for specimens of various sizes and spans tested under third-point loading with longitudinal axis parallel or perpendicular to the length dimension of the sheet as produced.

For comparison with newer alternative composites the primary specification criteria for flat products are converted to flexural strengths measured longitudinally (parallel to the length of the sheet and parallel with the fiber lay) and transversely, and are tabulated with absorption limits (Table 6.1). Flexural strength requirements are the same in each direction for flat sheet products, while lower strengths are accepted in the weaker direction perpendicular to the lay of the fibers for roofing shingle and siding products. Generally, the flexural strength requirement decreases as the nominal product thickness increases.

Flexural strength and 24-hour water absorption are the only "routine" tests performed according to the prescribed testing procedures in ASTM C459. The ASTM C 220 specification for flat sheet, contains deflection criteria which are not routine for acceptance testing, but it is noted in ASTM C 459 that the greater the deflection at maximum load the lower the breakage hazard in use, provided of course the flexural strength is satisfactory. This notion that a minimum deflection is desirable in conjunction with meeting the strength requirement is consistent with the idea that ductility and toughness, rather than strength alone, are important in the applications for which asbestos cement or newer alternative composites are suited.

6.2. SPECIFICATIONS FOR NON-ASBESTOS FIBER-CEMENT PRODUCTS

ASTM specifications for non-asbestos products are in a continual state of evolution that started in 1991 (ASTM, 1995). They include ASTM C 1186 (flat sheets Type A for exterior use such as claddings, facades, curtain wall, soffits etc., and Type B mainly for interior use such as partitions, floors and

TABLE 6.1 ASTM performance requirements for flat asbestos-cement products

ASTM standard specification	Flexural strengths-MPa for nominal thicknesses			Water absorption maximum — %
	4.5 mm	6 mm	10 mm	
C 220-Type F (flat sheet)	27	27	22	25%
C 220-Type U (flat sheet)	17	17	14	30%
C 222 (roofing shingle)	17 14*	17 14*	12 10*	25%
C 223 (siding)	17 14*	17 14*	12 10*	30%

* Transverse direction perpendicular to lay of fibers.

ceilings) and C 1225 (roofing shingles, shakes and slates). These specifications embody flexural strength requirements based on centre-point loading of specimens in a water-saturated condition (48 hr in 23°C water) or a standard "equilibrium" moisture condition (4 to 7 days, depending on thickness, at 23°C and 50% relative humidity). Flexural strength is measured on corresponding pairs of specimens cut parallel and perpendicular to the length of the sheet as produced, and the average reported. Four grades of strength are permitted for flat sheets specified to ASTM C 1186. Only one strength grade is specified for roofing products in ASTM C 1225 (Table 6.2).

Both standards also provide for supplementary testing to assess moisture content, water tightness and durability for various exposure regimes defined in the test method ASTM C 1185 (Table 6.3). Numbers are reported where applicable (NR), but there are no numerical requirements. Water tightness is judged satisfactory if water drops (WD) do not form during the test. Durability with respect to frost, warm water, and heat/rain resistance is judged satisfactory if there is no evidence of visible cracks (VC) or structural alteration likely to affect performance in use. Strength ratios (SR) based on final

TABLE 6.2   ASTM strength requirements for flat non-asbestos cement products

ASTM standard specification		Flexural strengths — MPa	
		Wet	Equilibrium[a]
C 1186	Grade I	4	4
(Type A flat	Grade II	7	10
sheets for	Grade III	13	16
exterior use)	Grade IV	18	22
C 1225 (roofing products)		5.5[b]	–

[a] Values also apply to Type B sheets for interior use.
[b] Primary strength direction. Not less than 50% of this value in weakest direction.

strength after test divided by initial saturated strength are reported for freeze/thaw and warm water testing regimes, but there are no numerical minimum requirements at present (Table 6.3). A 1996 proposed requirement for roofing products calls for 75% flexural strength retention after 100 cycles of freezing and thawing.

Both standards are intended to cover a wide range of fiber-reinforced cements, defined quite generally in terms of composition as "consisting essentially of an inorganic hydraulic cement binder or a calcium silicate material with a calcareous material reinforced by organic fibers, inorganic non-asbestos fibers or both" with appropriate process aids, fillers and pigments also permitted. They are under the jurisdiction of ASTM Committee C 17 and the requirements can be applied to most of the fiber-reinforced cements considered in the remainder of the chapter. However, the requirements for glass fiber-reinforced cement have evolved separately under the jurisdiction of ASTM Committee C 27.

6.3. GLASS FIBER-REINFORCED CEMENTS (GFRC)

Following the recognition that the glass fibers developed for use in fiber-reinforced plastics in the 1950's, borosilicate E-glass and soda-lime-silica A-glass, deteriorated rapidly in the highly alkaline environment of cement paste, a zirconia-based alkali-resistant (AR) glass was developed by the Building Research Establishment and Pilkington Brothers in the U.K. and marketed as multifilament strand in 1971. Other similar glass strand products followed from Owens-Corning in the U.S.A. and Nippon Electric Glass in Japan, all characterized by 16–20% zirconium oxide as the key constituent (ACI Committee 544, 1996).

6.3.1. AR Glass-Cement Interaction

Unfortunately, even AR glass strand in GFRC composites is subject to embrittlement in cement for reasons that involve

TABLE 6.3 Supplementary requirements for flat non-asbestos cement products

Supplementary test procedure	ASTM C 1186		ASTM C 1225	Results
	Type A	Type B		
Moisture movement (length change from 30% to 90% R.H.)	✓	✓	✗	NR
Water absorption on saturation	✓	✓	✗	NR
Equilibrium (50% R.H.) moisture content	✓	✓	✗	NR
Water tightness (50 mm water head for 24 hours)	✓	✗	✓	WD
Frost resistance saturated (–20°C to 20°C for 50 cycles)	✓	✗	✓	VC
Warm (60°C) water immersion for 56 days	✓	✗	✓	VC,SR
Heat (60°C) and rain (30°C spray) for 25 cycles	✓	✗	✓	VC,SR

✓ – required, ✗ – not required.

more than simply alkali attack of the fibers. Indeed, direct alkali attack of the glass fiber surfaces is probably not the dominant cause of deterioration for AR glass fibers because the rate of leaching of glass constituents from single filaments immersed in cement extract solution decreases sharply after about 14 days and proceeds only very slowly thereafter (Majumdar and Laws, 1991). It may even cease eventually due to a greater concentration of the relatively invulnerable zirconium constituent at the filament surface protecting the underlying more vulnerable silicon constituent after the early-age depletion of the surface silicon, essentially eliminating later-age decrease in filament strength (Majumdar and Laws, 1991).

The second and most widely accepted reason for the embrittlement of AR-GFRC composites, which incorporate multiple filaments coated with a sizing, is that calcium hydroxide crystals formed from cement hydration reactions intrude into the interstitial spaces between the individual filaments, and bond increasingly to the strand as hydration proceeds. The gradual increase in bond with age adversely affects the tensile breaking strength of the strand in strand-in-cement tests (Fig. 6.1) (Majumdar and Laws, 1991) (Proctor, 1986) which have proven to be a convenient means of evaluating the severity of the interaction between glass strand and matrix (Litherland, Maguire and Proctor, 1984). Considerable effort has been directed towards ameliorating this dominant cause of embrittlement in GFRC composites, including a comprehensive research program that evaluated well over 100 glass compositions (Fyles, Litherland and Proctor, 1986).

North American requirements developed by the Precast/Prestressed Concrete Institute (PCI) reflect the importance of zirconia content (Fig. 6.1) and the use of the strand-in-cement test to assess the severity of the strand-cement interaction. They are contained in a specification requiring a minimum zirconia content of 16% for the glass and a minimum retained strength of 330 MPa for strand-in-cement tests after 96 hours immersion in water at 80°C (PCI

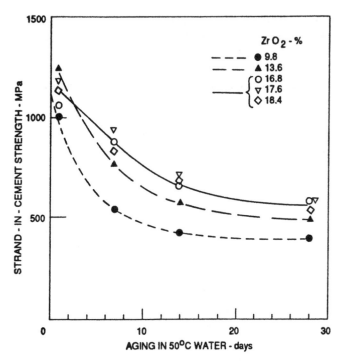

FIGURE 6.1   Effect of zirconia content on aging of glass strand in a
water-saturated cement matrix (Majumdar and Laws,
1991)

Committee, 1991, Appendix D). Commercially available AR
glass strands vary in zirconia content from 17% to 20% and
in the amounts and types of other oxides present (ACI
Committee 544, 1996).

In the continuing search for improving the long-term performance of GFRC, optimizing glass composition is only one of
several approaches for addressing the problem of embrittlement with aging. Modifying the matrix by use of pozzolans to
consume the calcium hydroxide (lime) produced by normal
portland cements so that there is less of it available to infiltrate

the glass strand, or replacing portland cement with a low-lime or no-lime cement are other alternatives (Majumdar and Laws 1991) (Fyles, Litherland and Proctor, 1986).

6.3.2. Early-Age Properties

Flexural performance is the primary criterion of quality for thin-section GFRC products. It is evaluated according to ASTM C 947 using water-saturated test specimens of span-depth ratio between 16 and 30 cut from panels representative of the production process and subjected to third-point loading. Flexural strength at maximum load, commonly termed modulus of rupture (MOR) is reported, and deflection measurements by crosshead movement are used to establish the load-deflection relationship and specifically the point at which the initially straight portion of the relationship first departs from linearity. The stress corresponding to the load at this point using the elastic bending formula is reported as the proportional elastic limit (PEL). Other ASTM tests for quality control (ASTM, 1995) include dry and wet bulk densities, water absorption, and apparent porosity (ASTM C 948). In addition, glass fiber content can be monitored for as-produced material using a washout procedure on coupons cut from freshly formed uncured panels (ASTM C 1229). These and other tests relevant to quality control for GFRC are covered in a recast/Prestressed Concrete Institute manual (PCI Committee, 1991, Appendix F).

The early-age strength properties of spray-up GFRC made with a cement-based matrix and AR glass fibers show the expected dependence on fiber content and strand length (reflecting aspect ratio), as illustrated previously for direct tension (Fig. 4.2). Flexural strength (MOR) increases sharply with increase in fiber content up to about 6% fiber by volume, approximately 7.5% by weight (Fig. 6.2) (Majumdar and Laws, 1991). Beyond this point strength levels off or decreases, probably due to incomplete wetting of the strands by the matrix and poor compaction of the composite after

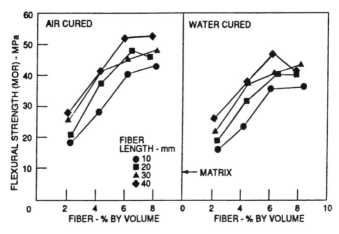

FIGURE 6.2   Early-age flexural strength of spray-up GFRC after
curing for 28 days (Majumdar and Laws, 1991)

spraying, as indicated by corresponding densities (Fig. 6.3).
However, impact resistance does not apparently peak at the
6% fiber content, and is also less sensitive to change in strand
length or aspect ratio than MOR (Fig. 6.4). Generally, strength
and especially impact resistance are slightly higher for air-
cured samples than for water-cured samples (Fig. 6.2 and 6.4),
possibly because cement hydration and consequent lime-asso-
ciated strand embrittlement are more advanced after 28 days
of water curing.

For vibration-molded premix GFRC, the patterns of
increase in 28-day strength and impact resistance with increasing fiber content are similar to those for spray-up material
(Fig. 6.2 and 6.4), but the values at any give fiber content are
lower (Fig. 6.5) (Majumdar and Laws, 1991).

In North America most GFRC is produced by the spray-up
process to meet the requirements of the Precast/Prestressed
Concrete Institute recommended practice for GFRC panels
(PCI Committee, 1993) which stipulates a minimum thickness
of 13 mm and a minimum fiber content of 4% by weight. For
the 5% fiber by weight (approximately 4% by volume) typical

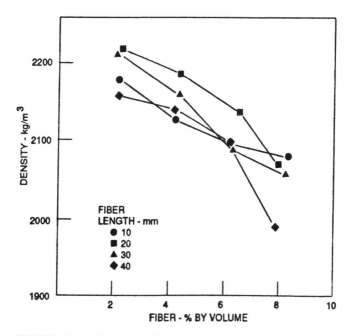

FIGURE 6.3   Influence of fiber amount on density of spray-up
GFRC at age 28 days (Majumdar and Laws, 1991)

of commercially produced spray-up GFRC, water-cured MOR
values are in the range 20–30 MPa (Fig. 6.2). Although there
is at present no ASTM performance specification for GFRC, a
specified minimum MOR of 18 MPa is apparently common in
the industry. This corresponds to the grade IV wet strength
requirement in the ASTM C 1186 specification for flat non-
asbestos sheet products (Table 6.2), although the mode of
loading (third-point vs. centre-point) and specimen dimensions
differ between the standard tests for GFRC (C 947) and flat
non-asbestos products (C 1185).

6.3.3. Aged Properties

Long-term natural weathering effects on GFRC and accelerated tests to simulate them over a shorter time span in warm

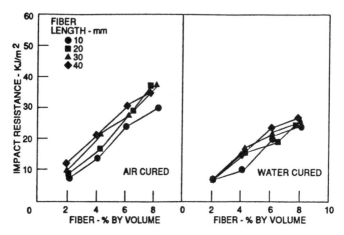

FIGURE 6.4   Early-age impact resistance of spray-up GFRC after
curing for 28 days (Majumdar and Laws, 1991)

water have been studied intensively using strand-in-cement
tests and water temperatures of 4°C to 80°C. Arrhenius plots
covering the time period for strand-in-cement strengths to
decrease from 1000 MPa to 300 MPa were used to equate days
in 50°C water to years of natural weathering for composite
flexural strength assuming correspondence between rate losses
for strand-in-cement strength and composite flexural strength
(Fig. 6.6) (Litherland and Proctor, 1986) (Majumdar and
Laws, 1991). For spray-up GFRC with normal portland
cement and 5% by weight of AR fibers the correlation for
flexural strength (MOR) shows that the results of accelerated
tests in warm water are related to results for natural weathering, in this case weathering in the U.K. climate (Fig. 6.7). For
example, 180 days in 50°C water is equivalent to about 50
years of typical U.K. weather with a mean annual temperature
of 10°C. Other qualitatively similar relationships can be
expected for different climates and composite constituents.

Obviously, the severity of the glass-cement interaction for
any particular application depends on the actual weathering

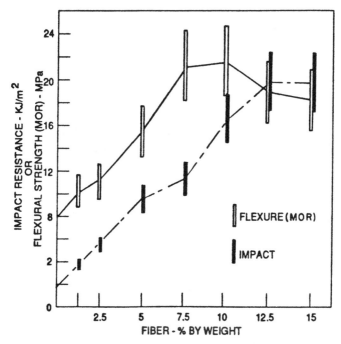

FIGURE 6.5   Early-age flexural strength and impact resistance of premix GFRC after curing for 28 days (Majumdar and Laws, 1991)

environment with respect to both temperature and humidity. It is greatest in warm moist exposure conditions and least in cool dry conditions. However, qualification testing for predicting the potential long-term effects of aging makes the adoption of a standardized accelerated aging regime desirable. The PCI test (PCI Committee, 1991, Appendix F) is based on aging of flexural test specimens in water at 50°C or 60°C. The time after which a doubling of the aging period in water causes less than a 5% decrease in strength or 10% decrease in strain capacity is reported as the retention time at which the aging process can be considered complete. The aged flexural

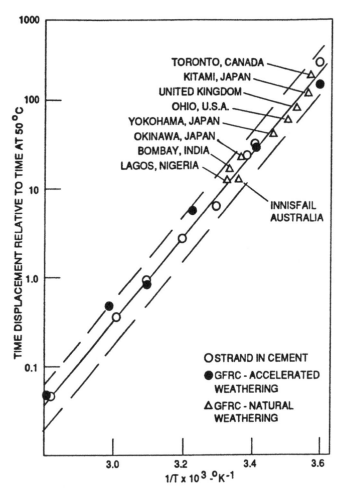

FIGURE 6.6   Rates of loss of strand-in-cement strength and GFRC
composite flexural strength in accelerated aging and
natural weathering relative to aging in 50°C water
(Litherland and Proctor, 1986)

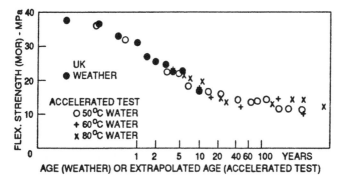

FIGURE 6.7   Effect of natural and accelerated aging on the strength
of GFRC (Majumdar and Laws, 1991)

strength and strain capacity are reported, but no minimum performance requirements are presently specified by either PCI or ASTM, although an ASTM standard is being developed.

In earlier work (Litherland and Proctor, 1986) and in the PCI accelerated aging test, the thrust has been to establish the severity of the loss of flexural strength and the time span over which it occurs. However, the reality is that under severe conditions unmodified GFRC composites suffer substantial embrittlement and consequent reductions not only in strength (Fig. 6.7) but also in strain capacity (Fig. 6.8) and toughness (Fig. 6.9) (Shah et al., 1988). Modifications to prolong the long-term retention of these properties are a high priority in ongoing research.

6.3.3.1. Modification to the Glass Strand

One approach employs incorporating a chemical inhibitor into the sizing on the glass strand to slow the rate of growth of lime crystals between the filaments, as in Cem-FIL 2 developed by Pilkington Bros. in the U.K. and NEG H200 developed by Nippon Electric Glass Inc.. This improves retention of both strand-in-cement strength (Fig. 6.10) (Majumdar and Laws, 1991) and composite flexural strength in natural and acceler-

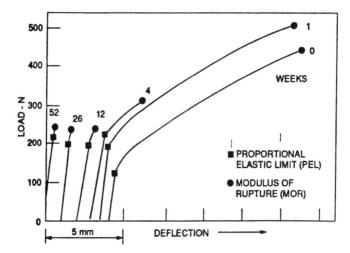

FIGURE 6.8   Effect of 1–52 weeks of aging in 50°C water on the
flexural stress-strain behaviour of GFRC (Shah et al.,
1988)

ated weathering tests (Fig. 6.11) (Litherland and Proctor,
1986) (Litherland, 1986).

6.3.3.2. Polymer Addition

Another approach involves adding a water-dispersed polymer
during the mixing process. According to Bijen (1990) it has
three effects. It decreases porosity and absorption in the composite thereby reducing the space for free water to exist and
promote transport of lime, helps to fill the space between glass
filaments available for deposition of lime, and forms a protec-
tive surface film on the strand. Seven-year tests in Europe
using a pure acrylic polymer (Forton VF 774) with a portland
cement and 5% by weight of AR glass strand showed the
polymer-modified GFRC with 15% polymer by weight of
cement performed better than the unmodified GFRC under
natural Dutch weathering conditions. However, for continuous
immersion in 20°C water the effect of the polymer was

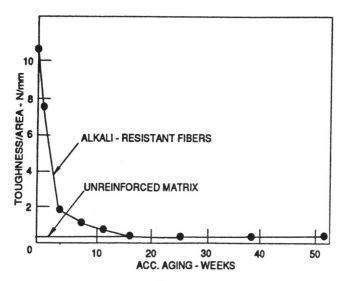

FIGURE 6.9   Effect of aging in 50°C water on toughness of GFRC
in terms of area under the load-deflection relationship
(Shah et al., 1988)

minimal after 7 years, and losses of flexural strength and
impact resistance were about equal for both the polymer-
modified and unmodified composites (Fig. 6.12) (Bijen, 1990).

Two forms of accelerated test were performed, one involving immersion in 50°C water (consistent with PCI practice in
North America), and the other involving cycles of wetting and
drying (20°C water for 24 hours followed by forced air drying
at 70°C). The wet/dry cycling test results correlated with
natural weathering results better than the results for hot water
immersion, so it was accordingly argued that the hot water
immersion test is unrepresentative of the natural Dutch weathering conditions applicable in this case (Bijen, 1990), imply-
ing that the test is too severe.

Nevertheless, North American practice tends to highlight
immersion in 50–60°C water as the criterion for assessing
long-term durability (PCI Committee, 1991, Appendix F).

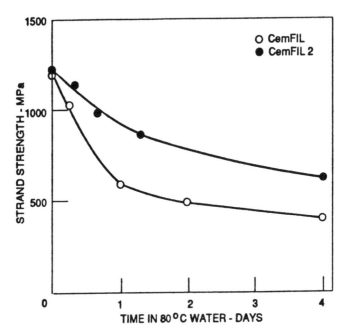

FIGURE 6.10   Effect of chemical inhibitor on strand-in-cement
              strength for aging in water at 80°C (Majumdar and
              Laws, 1991)

Results using 6% fibers by weight with an acrylic polymer and
immersion in 50°C water for 50 days show improvements in
unaged flexural strength (MOR) and toughness especially at
15% by polymer weight of cement (Fig. 6.13) (Soroushian et
al., 1993). However, while polymer-modified GFRC performed better than the unmodified form after accelerated aging
in 50°C water, the polymer did not prevent significant losses in
both strength and toughness (Fig. 6.13). Losses in toughness
were more severe than losses in flexural strength, with over
80% loss in toughness after 7 weeks aging in 50°C water, consistent with the trend in Fig. 6.9.

    Subsequent work (Bijen, de Haas and Bosmans, 1994) (Ball
and Wackers, 1993) established the combined effects of

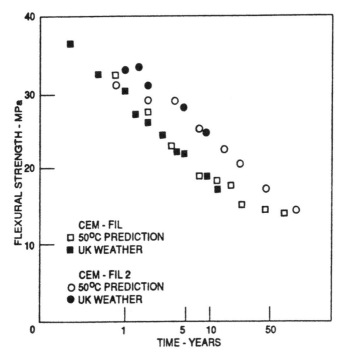

FIGURE 6.11   Effect of chemical inhibitor on GFRC strengths for natural and accelerated weathering (Litherland and Proctor, 1986)

polymer modification and use of glass strands coated with chemical inhibitor, Cem-FIL2 and NEG H200, along with comparative data for the original Cem-FIL strand. Fiber content was 5% by weight in a matrix of sand-cement ratio 1.0 with polymer contents of 0, 7, 10 and 15% Forton VF 774. Properties after accelerated aging for 12 weeks in 50°C water, or 100 cycles of wetting and drying (24 hours in 20°C water followed by forced air drying for 24 hours at 70°C) were compared with values established prior to the start of accelerated aging tests. The unaged values are based on 28 days of curing at 20°C and 65% relative humidity for the polymer-modified GFRC and moist curing for the first 7 days followed by

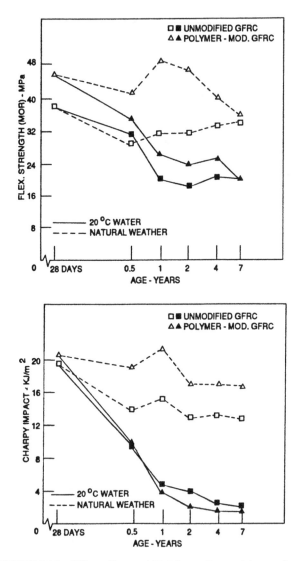

FIGURE 6.12   Effect of natural Dutch weather and immersion in 20°C water on flexural performance of polymer-modified and unmodified GFRC in terms of flexural strength (upper) and impact resistance (lower) (Bijen, 1990)

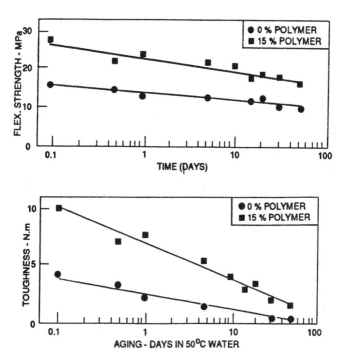

FIGURE 6.13   Effect of polymer modification on aging of GFRC in
50°C water in terms of flexural strength (upper) and
toughness (lower) (Soroushian et al., 1993)

21 days at 65% R.H. for the unmodified GFRC. The results
for 0, 7 and 15% polymer content once again show that losses
in strain, and therefore toughness, are more severe than corresponding losses in strength (Fig. 6.14). Polymer treatment
enhances both the unaged strength and the strength retained
after aging in 50°C water, with strength retention about
70–75%. It appears rather more effective in improving the
strength after aging in the European cyclic wetting and drying
test. Strain retention after aging in 50°C water or cyclic
wetting and drying is also improved by polymer treatment,
apparently more for the Cem-FIL2 and NEG H200 strands

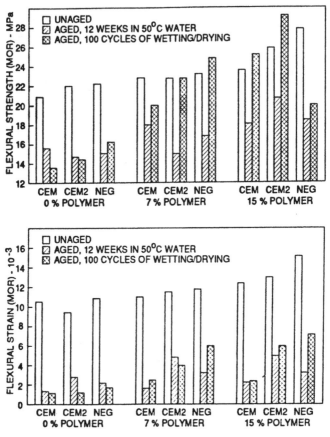

FIGURE 6.14   Effect of polymer modification and type of glass
strand on aging of GFRC in 50°C water in terms of
flexural strength (upper) and strain capacity (lower)
(Bijen et al., 1994)

than for the original Cem-FIL, but is still only 20–40% even
for these improved forms of glass strand when aged in 50°C
water.

Since the design philosophy for GFRC panels (PCI
Committee, 1993) is based on ensuring that design stresses

remain below the 28-day proportional elastic limit (PEL) which is largely a matrix-dependent property essentially independent of the glass-cement interaction, it is worth noting that polymer treatment may slightly improve the PEL retained after aging (Bijen, de Haas and Bosmans, 1994). Some of this improvement is attributable simply to the beneficial influence of warmth on hydration of the matrix, 50°C or 70°C during either of the aging tests. In the cyclic wetting and drying regime the polymer also has the beneficial effect of inhibiting moisture movement into or out of the matrix. Consequently, dimensional changes associated with wetting and drying are reduced and dimensional stability improved. Water absorption is also reduced (Bijen, de Haas and Bosmans, 1994).

Undoubtedly, the most important role of polymer additives is to facilitate curing by eliminating the need for moist curing following manufacture. In this regard, the specification for polymer curing agents (PCI Committee, 1993, Appendix L) requires that polymer-modified GFRC with no moist curing shall achieve a 28-day flexural strength at least equal to the 28-day strength of unmodified GFRC cured moist for the first 7 days. In addition, the polymer-modified dry-cured GFRC shall have MOR, PEL and strain capacity at least equal to the unmodified GFRC with 7 days moist curing when they are subjected to the prescribed aging test in 50°C or 60°C water (PCI Committee, 1991, Appendix F).

6.3.3.3. Matrix Modification

Replacing portland cements with their high lime content as a hydration product in the matrix by an alternative cement that generates little or no lime is one approach. High alumina cement (HAC) and several supersulphated cements (SSC) based on blending slag with calcium sulphate and a small amount of lime or portland cement as an activator were evaluated in U.K. work. However, each had disadvantages. Conversion weakens the matrix under warm moist conditions for HAC, and carbonation under atmospheric weathering that destroys ettringite formed during hydration weakens the matrix for SSC (Majumdar and Laws, 1991).

More recently, another form of slag-based low-alkali cement containing calcium sulphoaluminate, called CGC cement, was evaluated in Japan and showed promising results with AR glass fibers when immersed in 70°C water (Hayashi et al., 1986). Tests using this cement with 5% by weight of NEG glass strand showed that retention of flexural strength (MOR) and strain at MOR are much better than for unmodified GFRC under conditions of immersion in 80°C water (Fig. 6.15) (Akihama, Suenaga, Tanaka and Hayashi, 1987).

In North America, rapid hardening hydraulic cements based on portland cement with calcium sulphoaluminate have been used in a new matrix system (Zircrete) that includes a retarder to control setting, a plasticizer to enhance workability and a pozzolanic additive to combine with any residual lime left after the primary hydration reaction is complete (Molloy, Jones and Harmon, 1994). It is prepared and cured without moisture or polymer, as described previously in 5.1.2. This GFRC system gains strength relatively rapidly compared with unmodified GFRC, reaching a flexural strength (MOR) of the order of 19 MPa at 24 hours which is over 80% of its strength at 28 days (Molloy and Jones, 1993). After aging up to 100 days in 60°C water, there is virtually no loss of flexural strength, and the rate of loss of strain is much slower than for comparable unmodified GFRC, with over 70% of the original strain capacity retained after 100 days (Fig. 6.16) (Molloy, Jones and Harmon, 1994). Initial drying shrinkage is also reported to be less than for unmodified GFRC, but the choice of water-cement ratio, recommended between 0.43 and 0.50, and water-curing to prevent temperature rise during the initial set and exothermic reaction are important. Too little or too much water can increase shrinkage to a level similar to that for unmodified GFRC.

Another form of matrix modification involves using normal portland cement supplemented by a pozzolan to consume the

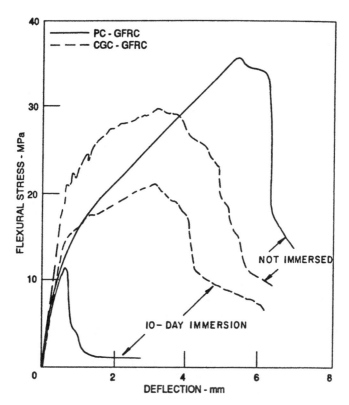

FIGURE 6.15   Effect of CGC cement on aging of GFRC in 80°C water in terms of flexural strength and strain capacity (Akihama et al., 1987)

lime formed during hydration that is the main cause of adverse glass-cement interaction. However, a highly reactive pozzolan is needed to maximize its effectiveness in inhibiting the interaction, so recent emphasis has been on silica fume and metakaolinite rather than on fly ash. Moreover, mixing the pozzolan with the rest of the matrix ingredients is not as effective, at least for silica fume, as pretreating the glass strand by dipping it in a silica fume slurry prior to mixing with other

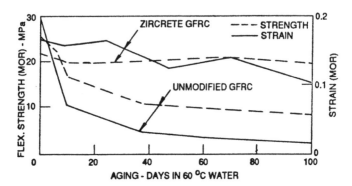

FIGURE 6.16   Comparative flexural strength and strain retention of
               Zircrete and unmodified GFRC after aging in 60°C
               water (Molloy, Jones and Harmon, 1994)

matrix ingredients (Bentur, 1990). Clearly, this gets more of
the active pozzolan close to the scene of the glass-cement
interaction. Accordingly, for a given amount (and cost) of
silica fume, the pretreatment approach is techically more
effective. There is little or no loss of strength after aging in
50°C water for 5 months, and the loss of toughness is considerably less than for unmodified GFRC and appears to stabilize
after 2 months (Bentur and Diamond, 1987). However, the
slurry pretreatment process may be impractical because of
associated increases in the cost and complexity of production.

Yet another variation of matrix modification with a pozzolan
involves the addition of 20–25% metakaolinite by weight of
cement to the matrix ingredients at the time of mixing (PCI
Committee, 1993). This additive is highly reactive, thus producing high early-age strengths while suppressing the glass-
cement interaction by consuming lime. The minimum amount
needed depends on the reactivity of the particular metakaolinite selected. It is evaluated in a test that determines its
lime-consuming potential, and relates the result to the amount
of metakaolinite needed to consume all the lime produced by
cement hydration (Thiery, Vautrin and Francois-Brazier,

1991). It may be used with or without a polymer additive to augment protection of the strand and eliminate moist curing.

In cyclic wetting/drying tests using the European procedure (immersion in 20°C water for 24 hours followed by forced air drying at 70°C for 24 hours), the effect of metakaolinite at 25% by weight of cement was to slightly improve flexural strength retention and significantly improve strain retention in mixtures with 5% by weight of AR glass and a sand-cement ratio of 0.70 (Fig. 6.17) (van der Plas, Yue and Bijen, 1992). However, specimens with metakaolinite and no polymer were found to undergo significant surface crazing during the wetting

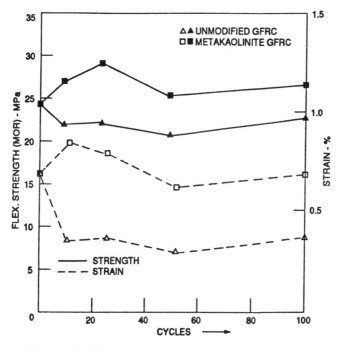

FIGURE 6.17   Effect of metakaolinite on flexural strength and
strain retention in GFRC subjected to cyclic wetting
and drying (van der Plas, Yue and Bijen 1992)

and drying to 100 cycles. Addition of an acrylic copolymer can effectively prevent such crazing, but the effects of the metakaolinite and the polymer are not cumulative as the levels of strength and strain achieved with polymer differ little from those with the metakaolinite alone. Naturally, the polymer reduces absorption and dimensional change associated with wetting and drying, as discussed previously for polymer-modified GFRC without metakaolinite in 6.3.3.2.. Significantly improved strain and toughness retention characteristics are also reported for polymer-modified metakaolinite GFRC immersed in 50°C water for up to 84 days (Thiery, Vautrin and Francois-Brazier, 1991).

6.3.3.4. GFRC Design Philosophy

While many attempts have been made to overcome the intrinsic chemical incompatibility of glass fibers with portland or any other lime-bearing cements, long-term durability continues to be one of the most important issues governing the use of GFRC in a diversity of exposure conditions that can range from the most severe, warm and moist, to the least severe, cool and dry. Some of the modified GFRC's discussed appear to offer the promise of less severe losses in strength and ductility with time than for unmodified GFRC made simply with a portland-cement based matrix and AR-glass fibers. However, the current N.American design philosophy (PCI Committee, 1993) continues to reflect the worst-case scenario where long-term properties eventually regress to values approaching the composite PEL, with allowable design stresses based on a specified fraction of the PEL after aging, the intent being to keep actual stresses below the matrix cracking strength at all times. As modified forms of GFRC evolve that unquestionably demonstrate sustained long-term strength and strain capacity much greater than the PEL values under severe exposure conditions, such as aging in 50°C water, the design philosophy may evolve towards permitting higher design stresses.

6.3.3.5. GFRC Applications

The early applications explored in the 1970's at the U.K. Building Research Establishment (BRE) included cladding panels for the exterior of buildings, window frames, pipes and permanent formwork (Ryder, 1975). A variety of GFRC spray-up products were produced at the BRE in 1970's (Fig. 6.18). Because of concern about reductions in strength, toughness and impact resistance with time in service, the BRE recommended that GFRC not be used in a primary load-bearing

FIGURE 6.18   GFRC products at the U.K. Building Research Establishment in 1974

structural capacity, and it compiled a digest of permissible and prohibited applications (Building Research Establishment, 1988).

In North America, a 1977 compilation of GFRC applications (Jones and Lutz, 1977) presents many photographs to support their view that the dominant application at the time in most countries was in architectural cladding panels. Both single-skin panels and sandwich panels (with thermal insulation included) prepared by the spray process continue to be widely used as the cladding for building facades (Majumdar and Laws, 1991). One of the most important ways of erecting these facades is the stud-frame construction technique developed in the U.S.A. that involves fabricating single-skin cladding panels integral with a steel frame that is subsequently attached to the building exterior (Fig. 6.19). The Precast/ Prestressed Concrete Institute has developed a recommended practice for the design of these integrated stud-frame GFRC panels (PCI Committee, 1993).

Design considerations for facades along with problems encountered and lessons learned from evaluation of in-place GFRC facades up to 1990 are discussed in a paper sponsored by the Portland Cement Association (Oesterle, Schultz and Glikin, 1990). Manufacture and installation experiences with GFRC facades are discussed in another PCA-sponsored paper in the same publication (Hanson, Roller, Daniel and Weinmann, 1990). Paramount in the conclusions are that aging must be adequately taken into account in design and that the consequences of GFRC embrittlement with loss of much of the glass fiber-reinforcing effect at some future time, often within the intended life span, must be recognized. The time for this to happen suggested for unmodified GFRC may be as short as 5 to 8 years in a warm humid climate and 20 to 30 years in a cooler drier climate.

Other applications for spray-up GFRC include permanent (left in place) or temporary formwork for bridge works, retaining walls, sewer linings, permanent collar units, utility boxes,

FIGURE 6.19   Steel framed architectural panel being demolded
                (upper) and cladding panels used in the De Anza
                Centre, California (lower) (Courtesy of Nippon
                Electric Glass America Inc.)

spun water and sewer pipes (Fig. 6.20), and troughs, drainage channels, tanks etc. for agricultural use (Majumdar and Laws, 1991). The use of GFRC as lining to rock face in mines and tunnels has also been explored (Greig, 1990), as have uses as artificial rock formations in landscaping (Fig. 6.21).

Premix GFRC is also being used for a variety of cast-in-place applications such as transformer pads, utility boxes, drainage channels and sunscreen panels (Fig. 6.22).

Another application for GFRC is as a plasticized or sprayed coating in surface-bonded masonry, where masonry blocks dry-stacked, with mortar used only to the minimum necessary to keep them plumb, are coated on both vertical faces with a hand-plastered or sprayed layer of GFRC. ASTM C 887 (ASTM, 1995) is a specification for surface bonding mortar made with glass fiber reinforcement and hydraulic cement with or without fillers. The standard also prescribes procedures for applicable tests including mixture consistency, time of set,

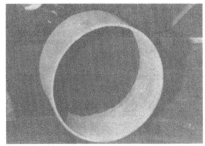

FIGURE 6.20   Spun GFRC pipe and GFRC used as formwork
(Courtesy of U.K. Building Research Establishment)

FIGURE 6.21   Artificial rock formation in GFRC (Courtesy of
Nippon Electrical Glass America Inc.)

water retention after mixing, and flexural and compressive
strengths after hardening. ASTM C 946 (ASTM, 1995) is a
standard practice covering workmanship and construction
procedures for dry-stacked, surface-bonded walls.

Nearly all products historically made of asbestos-cement,
such as general purpose flat sheet for buildings, roofing tiles or
slates, and even corrugated sheet, are being replaced with non-
asbestos alternatives, and GFRC is important for this purpose in
both North America and Europe (Majumdar and Laws, 1991).

6.4. CARBON FIBER-REINFORCED CEMENT
(CFRC)

Initial attempts to utilize carbon fibers in cement-based matrices involved high-strength, high-modulus fibers produced by
carburizing polyacrylonitrile (PAN) at high temperature. This
PAN-type of fiber proved to be very expensive, and interest

FIGURE 6.22   Premix GFRC for utility boxes, transformer pads, and sunscreen panels (Courtesy of Nippon Electric Glass America Inc.)

since the 1980's has centered mainly on the fibers produced from pitch extracted from petroleum or coal. These pitch-based carbon fibers are much less expensive than the PAN-type, although they are still expensive relative to many other fibers. The production technology for pitch-based fibers originated in Japan, and their tensile strength and elastic modulus, although lower than for the PAN-type, can vary over quite a wide range depending on the manufacturing process. Nishioka, Yamakawa and Shirakawa (1986) evaluated eight pitch-based types with properties in the range 445–764 MPa for tensile strength, 27–32 GPa for modulus of elasticity, and 1.7–2.4% for elongation. These are typical of the general purpose fibers made from isotropic pitch which produces a non-oriented fiber structure. Another stronger, stiffer, and less ductile category of pitch-based fibers is made from mesophase pitch which produces a highly oriented fiber structure with tensile strength 1500–3100 MPa, elastic modulus 150–480 GPa, and elongation 0.5–1.1% (ACI 544 Committee, 1996).

The properties of CFRC composites tend to reflect the fiber characteristics, with the stronger, stiffer, less ductile varieties giving relatively higher composite flexural strength (MOR) and relatively lower toughness compared with the weaker, less stiff and more ductile varieties, as in the comparison of PAN-type with pitch-based types (Fukushima et al., 1992).

Like other thin-section fiber cements, the properties of CFRC with any particular carbon fiber depend largely on fiber content and aspect ratio under the conditions of fiber pullout that normally prevail at failure. However, changes in factors that affect the fiber-matrix bond strength causing it to increase can produce more brittle behaviour because fiber fracture rather than pullout becomes dominant in the failure process, as discussed previously in 4.3.2 (Fig. 4.20 and 4.21). This means that even though the fiber content and aspect ratio remain constant, the fiber strength can become important in determining flexural strength and ductility (Fig. 6.23) (Nishioka, Yamakawa and Shirakawa, 1986).

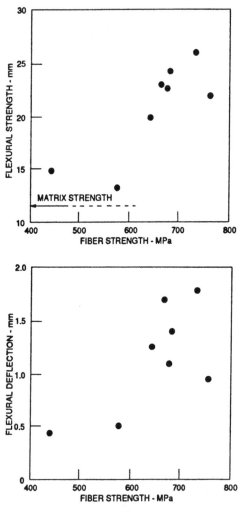

FIGURE 6.23   Effect of carbon fiber tensile strength on composite
performance in flexure with respect to strength
(upper) and ductility as indicated by deflection at
maximum load (lower) (Nishioka, Yamakawa and
Shirakawa, 1986)

The influence of aspect ratio is especially complicated for carbon fibers by their fragile nature and their consequent susceptibility to breakage during mixing, with the result that the aspect ratio of the fibers as they exist in the composite after mixing tends to be significantly lower on average than the initial value for the as-produced fiber prior to mixing. How much the aspect ratio is reduced by the mixing process probably depends on fiber length and bending strength, the size and amount of fine aggregate or fillers present, and certainly on mixing variables like mixer type, mixing time etc.. The reduction can be substantial even when using the relatively gentle Omni-mixer, as shown in the length distributions for 300–400 fibers after mixing where the average length is one quarter to one half the initial 10 mm length, with the lower strength, lower ductility fiber (440 MPa, 1.65% elongation) exhibiting more severe breakage than the stronger more ductile fiber (682 MPa, 2.33% elongation) (Fig. 6.24) (Nishioka, Yamakawa and Shirakawa, 1986). Consequently, composite strength reflects the actual fiber length or aspect ratio after mixing rather than the initial length or aspect ratio (Fig. 6.25).

Despite the susceptibility of these general purpose fibers to damage during mixing, they can be used to produce CFRC with substantial flexural strength depending on fiber content, fiber length and matrix type. For a very strong cement-based paste with 40% silica fume and 6% water-reducing admixture by weight of cement, flexural strengths reach about 28 MPa at 4% by volume of 3 mm long fibers (Fig. 6.26) with substantial toughness at 3% fiber and above (Fig. 6.27) (Ohama, Amano and Endo, 1985). Fibers of length 10 mm in this very strong autoclaved CFRC of water-cement ratio 0.30 performed not quite so well as the 3 mm length in all respects, probably due to fiber breakage in circumstances likely to promote very high fiber-matrix bond. For cement-based mortars of more moderate strength and water-cement ratio without silica fume, flexural strengths reach 20 MPa at 4% by volume of 10 mm fibers (Fig. 6.28) (Akihama, Suenaga and Banno, 1986–1).

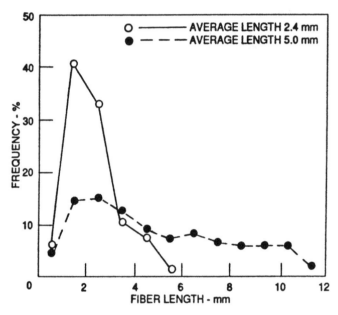

FIGURE 6.24   Distribution of carbon fiber lengths after mixing in
             an Omni-mixer for two types of carbon fiber
             of initial length nominally 10 mm (Nishioka,
             Yamakawa and Shirakawa, 1986)

However, inferior performance in terms of both flexural
strength (Fig. 6.28) and toughness (Fig. 4.21) was observed
when using a higher strength (w/c =0.30) paste matrix, again
probably due to the stronger interfacial bond strength causing
fiber fracture as already discussed. The need to select matrix
proportions that do not excessively increase interfacial bond,
thus promoting fiber fracture and inferior toughness in the
CFRC composite, has also been confirmed when using silica
fume in the matrix to more effectively disperse fibers as well
as to increase bond (Linton et al., 1991).

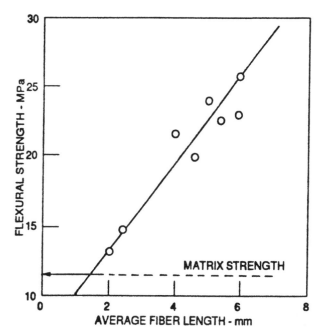

FIGURE 6.25   Effect of average fiber length after mixing on
flexural strength of CFRC (Nishioka, Yamakawa and
Shirakawa, 1986)

6.4.1. Modified CRFC's

Efforts to improve the basic CFRC system have included
polymer latex addition, modified cements to reduce length
changes caused by moisture movement, and the use of
stronger high-modulus fibers more resistant to breakage during
mixing and capable of producing improved flexural properties.
Polymer latex additions of 10% by weight of binder in a
cement-silica fume matrix with 3% by volume of fibers appear
to have little effect on flexural strength and impact resistance,
but tend to reduce water absorption and free shrinkage on
drying, and eliminate the need for moist curing (Soroushian,
Aouadi and Nagi, 1991).

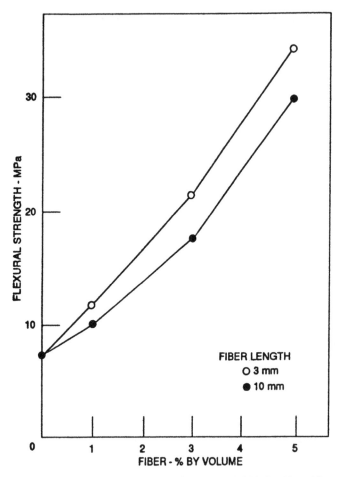

FIGURE 6.26   Flexural strength of autoclaved CFRC with a silica
fume-cement matrix (Ohama, Amano and Endo,
1985)

Higher strength (600–3000 MPa) and higher modulus
(30–240 GPa) fibers have been developed mainly to facilitate
production using a conventional mortar mixer, as described
previously in 5.1.3., instead of the special Omni-mixer (Ando

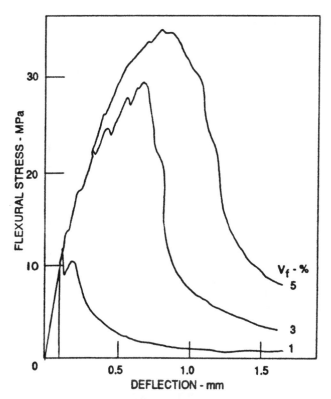

FIGURE 6.27   Flexural load-deflection relationships for autoclaved
CFRC with a silica fume-cement matrix (Ohama,
Amano and Endo, 1985)

et al., 1990). Flexural strengths using a mortar matrix of water-
cement ratio 0.45 correlate strongly with fiber strength, fiber
modulus, and fiber volume fraction, with the highest levels of
composite flexural strength at 4% fibers by volume reaching
20 MPa. The relationship between flexural strength and fiber
content for these fibers (Fig. 6.29) (Ando et al., 1990) is very
similar to that for general purpose fibers (Fig. 6.28) (Akihama,
Suenaga and Banno, 1986–1), despite the use of apparently

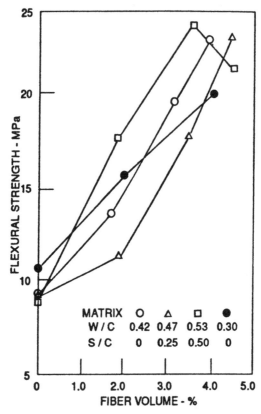

FIGURE 6.28   Flexural strength of CFRC pastes and mortars using
              an Omni-mixer with general purpose pitch-based
              fibers (Akihama, Suenaga and Banno, 1986–1)

stronger stiffer fibers, but the degree of fiber breakage as a
result of mixing was not discussed.

A shrinkage-reducing powdered admixture based on Al_2O_3,
CaO and SO_3 was used in one of three CFRC's containing
portland cement and silica powder or lightweight silica "bal-
loons" as aggregate (Yoda et al., 1992). The normal weight
composite without admixture reached flexural strengths of

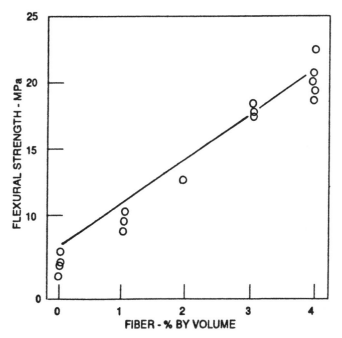

FIGURE 6.29   Flexural strength of CFRC mortars prepared using a
               conventional mixer with high-performance pitch-
               based fibers (Ando, et al., 1990)

about 10 MPa with 2% by volume of 3 mm fibers of aspect
ratio 167, considerably lower than the comparable values in
Fig. 6.28 for 2% of 10 mm fibers of aspect ratio 690, so the
importance of aspect ratio is again apparent. The lower aspect
ratio is also associated with much lower overall toughness.
The effect of the shrinkage-reducing admixture appears
marginal.

The apparently conflicting effects of fiber length or aspect
ratio in the data discussed are probably due to the fact that
only the initial fiber aspect ratio before mixing is known with
certainty, while it is the final fiber aspect ratio after mixing

that governs composite properties when fiber pull-out is dominant in the failure process. It is also apparent that when the fiber-matrix interfacial bond strength is too strong fiber breakage dominates the failure process, substantially reducing toughness. Accordingly, actual aspect ratios after mixing greater than the relatively low value needed to cause fiber breakage are of little benefit in a very strongly bonded system, while higher aspect ratios tend to be beneficial in a less strongly bonded system.

6.4.2. Durability

One of the least complex aspects of the behaviour of CFRC is durability. Unlike glass fibers, carbon fibers are stable in the highly alkaline environment of moist cement-based matrices, with no significant change in flexural strength or toughness after immersion in 75°C water for 5 months (Fig. 6.30) (Akihama, Suenaga and Nakagawa, 1988). Although exposed unprotected carbon fibers are attacked by strong oxidizing agents like sulphuric and nitric acid, the fibers in CFRC are protected by a cement-based matrix so its durability under acidic conditions depends mainly on the intrinsic resistance of the matrix to acid attack. In tests simulative of acidic weathering conditions, for example exposure to acid rain or sewage, no significant changes in composite flexural strength or toughness were observed after immersion in sulphuric or nitric acids of pH=4 for 90 days (Banthia and Sheng, 1991). However, the matrix in these tests was of extremely low permeabiilty due to a water-cement ratio of 0.30 and 20% silica fume by weight of cement, and its resistance to acid attack was enhanced by the silica fume. A higher water-cement ratio matrix without silica fume can be expected to deteriorate in acid conditions with consequent deterioration possible in the CFRC composite. However, polymer latex additives can help to reduce the severity of acid attack (Soroushian, Aouadi and Nagi, 1991).

The dimensional changes and possible cracking of CFRC subjected to cycles of wetting and drying tend to reduce with

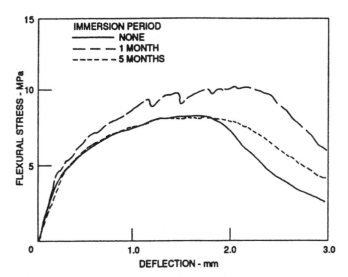

FIGURE 6.30   Flexural performance of CFRC immersed in 75°C
water (Akihama, Suenaga and Nakagawa, 1988)

increase in fiber content, in terms of both free shrinkage (Ohama, Amano and Endo, 1985) and cracking under restrained shrinkage (Soroushian, Nagi and Hsu, 1992). Unlike many other fiber-reinforced cements, the intrinsic stability of carbon fibers up to 300°C permits autoclave curing, typically at 150–180°C, and this in turn significantly reduces the shrinkage potential of the matrix and the corresponding composites (Akihama, Suenaga and Nakagawa, 1988).

6.4.3. Applications

In Japan, one of the main purposes of developing CFRC was to replace asbestos-cement in anticipation of restrictions or prohibitions on its use (Fukushima et al., 1992). CFRC is capable of easily meeting the flexural strength requirements of any of the four strength grades specified in ASTM C 1186 specification for flat non-asbestos cement sheets (Table 6.2). It is also likely that with appropriate choice of matrix proportions and production technique, it can meet the C 1186 requirements for supplementary weathering with respect to wetting and drying, warm water immersion and frost resistance. Satisfactory performance under Japanese climatic conditions is reported for a 15 mm thick cladding panel attached to the exterior of a building and for a 20 mm thick roofing panel to an outdoor corridor (Ando et al., 1990). Satisfactory interior use in free access floor panels, historically made from asbestos-cement, is also reported (Nishioka, Yamakawa and Shirakawa, 1986).

Several large-scale applications of autoclaved CFRC as cladding panels, curtain walls etc. are reported mainly in Japan. The first large-scale application was the use of precast, lightweight, autoclave-cured CFRC as cladding panels covering the 10,000 m^2 steel-formed domes that form the A1 Shaheed monument in Iraq. The panels were required to withstand the dimensional changes caused by the extremes of the Baghdad climate with very high temperatures and low humidities in summer and below-freezing temperatures in winter. The maximum panel loading of 60 kg/m^2 for the structure led to a specification requirement for the CFRC backing not to exceed 27 kg/m^2 in weight, and resulted in selection of a CFRC of specific gravity 1.0 and thickness 25 mm weighing 25 kg/m^2 topped with tiles weighing 33 kg/m^2. The CFRC was made using a special lightweight fine aggregate described as microballoons with 2% by volume of fibers in a superplasticized cement matrix of water-cement ratio 1.3 that achieved a flexural strength after autoclaving of 7.4 MPa at specific gravity 1.0. Five years after construction these curved tiled CFRC panels were reported to have performed satisfactorily (Fig. 6.31) (Akihama, Suenaga and Nakagawa, 1988).

Another large-scale application of CFRC is in the curtain wall units for the ARK high-rise office building in Tokyo. Precast, lightweight autoclave-cured panels of specific gravity 1.3 and area 4.55 m^2 were used for the 32,000 m^2 exterior of the building (Fig. 6.32) (Akihama, Suenaga and Nakagawa, 1988). Since autoclave curing restricts the size and types of

FIGURE 6.31   CFRC panels in the Al Shaheed monument in Iraq (Courtesy of Kajima Institute of Construction Technology)

finish possible in curtain wall panels, CFRC mixtures suitable for normal curing were developed, as described previously in 6.4.1, and appeared satisfactory after exposure to 6 months of outdoor weathering as 2x5 m curtain wall panels (Yoda et al., 1992). Other examples of satisfactory performance of CFRC curtain wall panels in large buildings after 2 to 4 years in service are reported (Sakai et al., 1994).

If production technologies evolve that reduce the relatively high cost of carbon fibers, the generally excellent strength, toughness and durability of CFRC composites will likely make them increasingly attractive for many thin-section applications.

6.5. ARAMID FIBER-REINFORCED CEMENTS

Aramid (aromatic polyamide) fibers are potentially attractive as fiber reinforcement because of their very high strength and

FIGURE 6.32   CFRC curtain wall panels for a high-rise office
building in Japan (Courtesy of Kajima Institute of
Construction Technology)

comparatively high elastic modulus. In cement pastes or
mortars prepared using an Omni-mixer, they appear capable
of substantially improving performance in terms of flexural
strength, toughness and impact resistance in accordance with
fiber content and aspect ratio (Fig. 6.33) (Akihama,
Nakagawa, Takada and Yamaguchi, 1986–2). They are capable
of withstanding autoclave curing at 140°C, and strengths after
autoclave curing and 14 days curing in air at 20°C and 65%
relative humidity are similar (Fig. 6.33, left). However, auto-

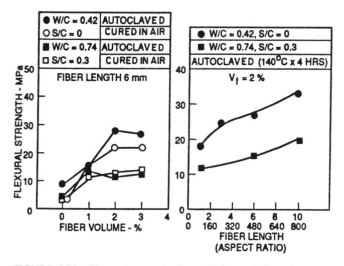

FIGURE 6.33   Flexural strength of aramid fiber-reinforced cements and mortars in terms of fiber content (left) and aspect ratio (right) (Akihama et al., 1986–2)

clave curing reduced shrinkage to 37% of the value for air curing. The levels of flexural strength reached (Fig. 6.33) are mostly within or above the 10–22 MPa range specified for flat non-asbestos cement products in ASTM C 1186 strength grades II, III and IV. In contrast, an attempt to incorporate aramid fibers into a silica fume-cement matrix using a conventional mortar mixer resulted in relatively low flexural strengths (3–5 MPa) and difficulty in obtaining adequate workability at fiber contents more than 1% by volume (Soroushian, Bayasi and Khan, 1990).

Although the relatively high cost of aramid fibers is probably the main impediment to their use, the compatibility of aramid fibers with cement-based matrices is also a concern. While satisfactory durability is reported for composites tested at ambient temperature for two years (Walton and Majumdar, 1978), a loss of flexural strength of about 10% after 6 months is evident for aramid fiber-reinforced mortars tested in 75°C water (Akihama et al., 1986-2). A similar loss of strength of aramid fibers aged in cement paste, especially at higher temperatures up to 50°C is also reported (Wang, Backer and Li, 1987). Furthermore, tests on uncoated multifilament aramid strand immersed in calcium hydroxide solution show a loss of strength that becomes increasingly severe with rising temperature, reaching almost 100% after 1 year at 95°C, while strand coated with epoxy resin is less severely affected (Schürhoff and Gerritse, 1986).

The relatively high cost and questionable durability of aramid fibers in cement-based matrices probably accounts for the lack of reports on engineering applications for this type of composite.

6.6. NATURAL FIBER-REINFORCED CEMENTS

In the developing countries many types of locally available natural vegetable fibers have been used to produce cement-based composites. There are at least two major compilations on the subject (Sobral, 1990) (Swamy, 1988). In these countries the preparation techniques are usually very basic without pressure compaction or dewatering, and yield comparatively low strength products intended for small building applications. The main attraction is the low cost of production associated with the abundance of locally available low-cost fibers and low-cost manual labor. The major disadvantage is the high water absorption of the fibers, their vulnerability to chemical attack by cement alkalis, and the detrimental effects of some secondary fiber constituents on setting and hardening of the matrix.

Direct chemical attack of the fibers occurs in several ways (Oliveira and Agopyan, 1992) (Gram, 1986). Cellulose molecular chains are oxidized and broken by alkalis, particularly at warm temperatures. Hemicellulose is similarly but more severely affected because of its lower degree of polymerization. Lignin which binds the cellulose-based fibers tends to dissolve in alkalis, particularly at warm temperatures. Accordingly, the rate of deterioration varies with fiber type depending on the proportions of cellulose, hemicellulose and lignin present in the natural fiber (Fig. 6.34) (Lewis and

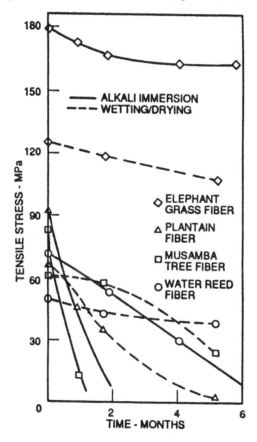

FIGURE 6.34   Loss of strength of natural fibers exposed to wetting and drying or immersion in calcium hydroxide solution (Lewis and Mirihagalia, 1979)

Mirihagalia, 1979). Of the three, the cellulose is most resistant to alkalis, and therefore treatments to reduce hemicellulose and lignin content are used to slow the alkali-related deterioration. They range from simply washing with water to reduce their presence at the fiber surface in contact with cement (Olivera and Agopyan, 1992) to sophisticated chemical kraft pulping in the developed countries. Reducing the alkali content by means of a locally available pozzolan such as rice husk ash, natural scoria or pumice can also help to improve long-term performance (Chatveera and Nimityongskul, 1992) (Berhane, 1994). Impregnation of fibers such as sisal with water-insoluble wood products has also been explored as a means of reducing attack by cement alkalis (Canovas, Selva and Kawiche, 1992).

The very high water absorption of most vegetable fibers is also a factor promoting weakness and deterioration in the fiber-matrix transition zone. The migration of water towards the fiber results in a water-cement ratio higher than in the surrounding matrix and a locally greater concentration of calcium hydroxide crystals, which in turn promote weakness under load due to easier debonding and consequent loss of strength and toughness (Savastano and Agopyan, 1992). Dimensional change caused by large moisture movements into or out of the fibers on wetting and drying combined with a tendency to rot under these conditions is also a concern (Fig. 6.34) (Lewis and Mirihagalia, 1979).

The leaching action of calcium hydroxide solution and even pure water is an additional factor that can detrimentally affect setting time and strength development in the matrix. Water-soluble constituents in the fiber that include hemicellulose, starch, sugar, tannins and lignins are leached by both water and calcium hydroxide solution. However, the amount of the resulting detrimental extract produced at any given time is several times greater for calcium hydroxide solution than for water, and the consequent retardation of setting and strength development more severe (Aggarwal and Singh, 1990).

Results for coir and sisal fibers show that the effect on set retardation is more severe for coir and on strength development more severe for sisal, so the compositional nature of the fiber is clearly important. The retardation effects of the extract can be reversed to some extent by use of an accelerating admixture (Aggarwal and Singh, 1990).

In general, the level of flexural strength achieved using natural vegetable fibers and simple fabrication techniques is quite low, often less than the 4 MPa specified as the lowest strength grade for flat sheet products in ASTM C 1186 and usually less than the 5.5 MPa required for roofing elements in ASTM C 1225, but the energy absorption capability or toughness is significantly improved. Both strength and toughness show deterioration with age under warm moist conditions for the reasons previously discussed, with the rate of deterioration dependent on the natural composition of the fiber (Fig. 6.35) (Savastano and Agopyan, 1992). Water-washed malva fibers have been used in Brazil to produce fiber-reinforced mortars having a 28-day strength of 7 MPa (Oliveira and Agopyan, 1992). In India, coir fiber-reinforced cement boards have been investigated for applications where they could be an alternative to increasingly scarce and expensive wood-based products like plywood and particle board (Aggarwal, 1992). Flexural strengths of about 10 MPa were achieved, and the boards performed satisfactorily in accelerated durability tests involving two different wetting and drying cycles or 200 hours immersion in boiling water. Accordingly, their use in panelling for door and window shutters, partitions, false ceilings, and cladding panels is being considered.

6.7. CELLULOSE-BASED FIBER-REINFORCED CEMENTS

Refined cellulose fibers less prone to alkali attack than their natural unprocessed equivalents can be produced by chemical

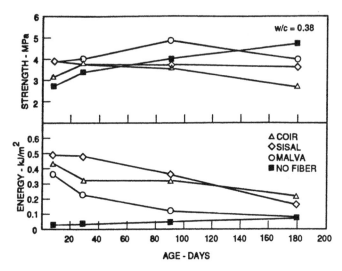

FIGURE 6.35   Deterioration of natural fiber-reinforced cement
composites under warm humid conditions in terms
of flexural strength (upper) and toughness (lower)
(Savastano and Agopyan, 1992)

or mechanical pulping of various species of wood and natural
vegetable fibers to reduce the more vulnerable lignin and
hemicellulose fractions. As indicated previously in 5.1.6.1,
the longer chemically pulped kraft softwood fibers are best
from the point of view of reinforcing effectiveness and resistance to attack by cement alkalis. However, a wide variety of
pulp-derived fibers from softwood, hardwood, hemp, cotton
and wastepaper, with a correspondingly wide range of chemical compositions and mechanical properties, have been investigated, and their performance in terms of composite flexural
strength related to the tensile strength of the parent fibers and
the degree of polymerization within them (Fig. 6.36) (de
Lhoneux and Avella, 1992). The longest and strongest fibers
provide the greatest composite strength, indicating that
composite failure involves at least some fiber fracture.

However, for fibers of the same strength those with the highest degree of polymerization produce the greatest composite strength, indicating that molecular fiber structure is also important. Flexural strengths are between 7 and 24 MPa (Fig. 6.36) using what appears to be a press and dewater manufacturing process similar in principle to the Hatschek process. A similar dependency of strength on fiber type and consequent composition is reported by Fordos and Tram (1986).

Autoclave curing may adversely affect strength development for some types of cellulose fiber by decreasing the degree of polymerization and tensile strength as well as releasing sugars that retard strength development in the matrix. This effect is least for highly refined pulps with minimal impurities where the fibers remain stable, in which case autoclaving can be beneficial by reducing free lime and lowering shrinkage potential (Fig. 6.37) (de Lhoneux and Avella, 1992).

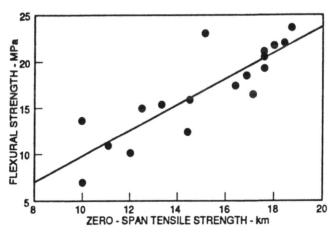

FIGURE 6.36   Relationship between flexural strength of pulp-based composites and the tensile strength of the fiber (measured as the length needed to produce fiber failure under its own weight) (de Lhoneux and Avella, 1992)

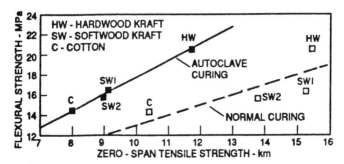

FIGURE 6.37   Effect of autoclave curing on relationship between
               composite strength and fiber strength for selected
               pulps (de Lhoneux and Avella, 1992)

Even within a single species wood, the reinforcing effectiveness of the fibers depends greatly on the production
process with respect to the proportion of summerwood
(stronger) to springwood (weaker) fibers and the intensity of
the mechanical processing used to break down larger fiber
groups into individual fibrils as described previously in 5.1.6.1
(Table 5.5) (Vinson and Daniel, 1990). Using 4–12% fiber by
weight flexural strengths were 8–10 MPa tested wet and 12–20
MPa tested dry for natural beater-refined and summerwood-
enriched unrefined fibers (Table 5.5), compared with 12–15
MPa tested wet and 25–30 MPa tested dry when the benefits
of both intensive beater refinement and summerwood enrichment were combined. These results also highlight the differences between wet and dry testing for normally cured (not
autoclaved) specimens, with strength reduced by wetting and
toughness in terms of area under the load-deflection relationship increased by wetting (Fig. 6.38).

Recycled wastepaper is another source of cellulose fibers, but
their effectiveness in improving strength and toughness depends
greatly on the composition of the original pulp and the extent to
which the recycling process has shortened the fibres and altered
their surface characteristics (Table 5.4) (Coutts, 1989). Flexural

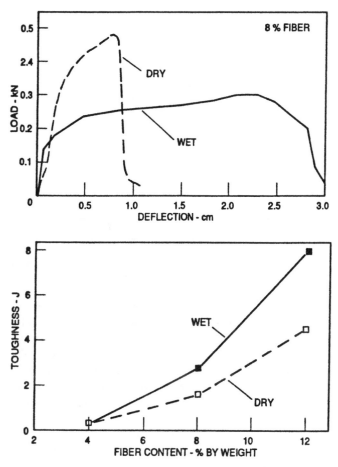

FIGURE 6.38   Effect of testing condition on flexural load-deflection
behavior (upper) and toughness (lower) for compos-
ites reinforced with a refined slash pine summerwood pulp (Vinson and Daniel, 1990)

strength and toughness are generally lower for the recycled
paper than for a premium pulp, and the effect of wet testing in
lowering strength is again evident (Fig. 6.39).

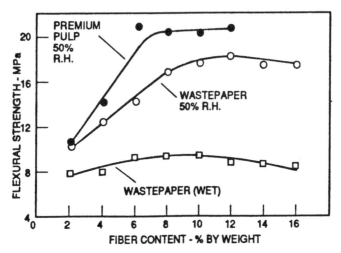

FIGURE 6.39   Flexural strength of composites reinforced with
wastepaper and premium pulp (Coutts, 1989)

Long-term durability is a concern for cellulose pulp-based composites, even when using the highly refined chemically produced kraft pulps that are largely free of lignin and are the preferred choice of fiber in the developed countries for the production of cellulose fiber-reinforced cement sheets in flat or corrugated form. There is general consensus that weathering leads to an increase in the fiber-matrix interfacial bond which promotes fiber fracture rather than pullout and therefore reduces toughness, while flexural strength is largely unaffected or may even increase (Akers and Studinka, 1989) (Bentur and Akers, 1989). Conditions of carbon dioxide enrichment accelerate the embrittlement due to more rapid depolymerization of the fibers, carbonation of the matrix, and other factors, while autoclaving reduces the rate at which embrittlement develops (Bentur and Akers, 1989). The minimal effect of weathering on strength and its more significant effect on toughness are compared for composites using 4% and 8% by weight of high quality softwood and hardwood kraft pulps tested according to the accelerated wetting/drying and warm water immersion

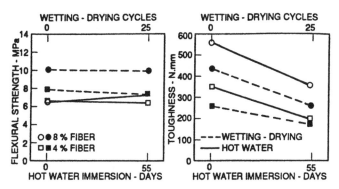

FIGURE 6.40   Flexural performance of normally cured wood fiber-reinforced cements after wetting and drying cycles or immersion in 60°C water (Marikunte and Soroushian, 1994)

procedures of ASTM C 1185 (Fig. 6.40) (Marikunte and Soroushian, 1994). Even in the presence of pozzolanic additives (30% fly ash or 15% silica fume by weight of cement), the loss of toughness was still significant for these non-autoclaved composites. Autoclaved composites made with paper pulp cellulose fibers show a similar tendency towards embrittlement in natural and accelerated weathering tests (Fig. 6.41) (Larsen and Krenchel, 1991).

Cellulose fiber-reinforced cements are generating considerable interest in North America for a wide range of flat and corrugated sheet products and roofing components meeting the requirements of ASTM C 1186 and C 1225 (ASTM, 1995), and have also been used for asbestos-cement replacement in roofing and facade applications in Switzerland, Germany, South Africa and South America (Studinka, 1989).

6.8. OTHER SYNTHETIC FIBER-REINFORCED CEMENTS

Other synthetic fibers considered for use in fiber-reinforced cements have included polyacrylonitrile (acrylic), polyvinyl

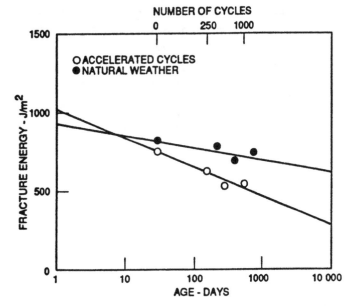

FIGURE 6.41   Deterioration in fracture energy (toughness) of autoclaved cellulose fiber-reinforced cement under natural and accelerated weathering (Larsen and Krenchel, 1991)

alcohol (PVA) and polypropylene (Studinka, 1989) (Gale, 1994) (Houget, Ambroise and Pera, 1995).

Acrylic fibers as primary reinforcement in a Hatschek manufacturing process with softwood kraft pulp and either polyethylene or polyoxymethylene pulps as the process fiber have been shown to produce composite flexural strengths of 10–15 MPa with good post-crack toughness characteristics in the short term (Daniel and Anderson, 1986) (Houget, Ambroise and Pera, 1995). Similar levels of strength are reported using conventional mixing to distribute up to 3% by volume of fibers (Odler, 1988). However, the long-term durability of acrylic fibers in cement is questionable because they are one of several fiber types (including polyester and aramid) that are reported as not resistant to strong alkalis (Lyle, 1976). Acrylic fiber strength losses were rated as slight for aging in cement paste at 22°C but significant after 60 days aging at 50°C and 100% relative humidity (Wang, Backer and Li, 1987). Accordingly, their ability to adequately withstand the ASTM C 1185 warm water immersion test (lime-saturated water at 60°C for 56 days) in a cement composite is doubtful. Nevertheless, acrylic fibers have apparently received approval for use in cement-based composites in Germany, despite evidence of a decrease of 8% in fiber strength after aging in 50°C water for 28 days (Hähne, Techen and Wörner, 1992).

PVA fibers as primary reinforcement in a Hatschek type of manufacturing process continue to attract interest in North America as an alternative to cellulose or glass fibers for non-asbestos products. Using 6% by weight of 15 mm long fibers in a superplasticized mechanically mixed mortar matrix, composite flexural strengths of 11–13 MPa were achieved along with near-plastic post-cracking behaviour sustained to large deflections (Houget, Ambroise and Pera, 1995). Using the Hatschek production process and an unstated but probably greater fiber content, unaged flexural strengths of 17–23 MPa were reported for a variety of proprietary products (Akers et al., 1989). Durability tests on the aged products after 3 to 7 years of natural weathering in Switzerland showed slight increases in flexural strength with time, attributed partly to matrix carbonation and an increase in fiber-matrix bond with aging. Accelerated aging tests involving 24-hour cycles of wetting at 20°C and drying at 80°C for 6 months also showed a slight increase in strength with time (Akers et al., 1989). Natural aging tests up to 4 years also show a slight increase in strength, but toughness decreased significantly by about 50% in the first year and then remained stable (Hikasa and Genba, 1986). Using an extrusion preparation process and 2–4% PVA fibers by volume, unaged flexural strengths of 17–27 MPa have been achieved with strain-hardening behavior and substantial ductility beyond the proportional elastic limit of

11–15 MPa (Shao, Marikunte and Shah, 1995) (Shao and Shah, 1997).

Polypropylene fibers as primary reinforcement for non-asbestos fiber cement products have been studied extensively in Denmark, but there is less interest in polypropylene for this type of application than in other alternatives such as cellulose, acrylic and polyvinyl alcohol (Gale, 1994) because of the lower levels of strength and toughness achievable. Nevertheless, composites made with polypropylene fibers can have useful flexural strengths of 6–9 MPa with reasonable post-crack strength and toughness (Houget et al., 1995). One major attribute of polypropylene is that, in contrast to cellulose, glass and many other alternatives, it is stable in the highly alkaline environment of cement paste, as confirmed by increasing composite toughness with age under natural weathering or accelerated weathering employing 12-hour wetting/drying and freezing/thawing cycles (Fig. 6.42) (Larsen and Krenchel, 1991).

A comparative performance evaluation of acrylic, polyvinyl alcohol and polypropylene as primary reinforcing fibers in cement sheets, produced using a slurry pressing and dewatering process with various cellulose-based pulps as a process fiber, shows flexural strengths with 3–5% fibers by weight in the range 15–22 MPa for both PVA and acrylic fibers, compared with only 10–15 MPa for fibrillated high bond polypropylene fibers which is not much different from the strengths using the process fibers alone (Table 6.4) (Simatupang and Lange, 1987). The satisfactory durability previously, reported for PVA reinforced composites is again confirmed by tests involving cycles of wetting and drying (20°C to 105°C) with freezing and thawing (–12°C to 20°C) which showed little change in dynamic modulus after 24 cycles. No comparable aging data are given for the acrylic or polypropylene fibers.

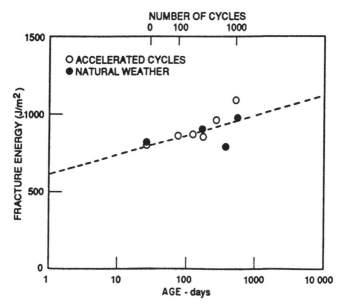

FIGURE 6.42 Increase in fracture energy (toughness) of polypropylene fiber-reinforced composites under natural and accelerated weathering (Larsen and Krenchel, 1991)

TABLE 6.4 Composite flexural strengths using 3% primary fiber with 7% cellulose-based process fiber pulp

Primary fiber	Flexural strengths – MPa			
	Pulp A	Pulp B	Pulp C	Average A,B and C
PVA (Kuralon)	15.2	17.2	18.3	16.9
Acrylic (Dolanit)	15.6	21.0	21.5	19.4
Acrylic (ATF1055)	13.6	16.3	18.3	16.1
Polypropylene (Krenit)	13.6	15.0	12.8	13.8
None	12.4	14.5	13.5	13.5

CHAPTER 7

Properties and Applications — Fiber-Reinforced Concretes

Fiber-reinforced concrete (FRC) is used mainly in relatively thick-section applications historically serviced by unreinforced or conventionally reinforced concrete. These include full-depth slabs on grade, overlays to existing slabs, and pneumatically applied shotcrete layers or linings used to stabilize, protect or rehabilitate exposed soil, rock, or deteriorated concrete surfaces, including surfaces exposed by tunneling. The minimum section thickness is usually at least 50 mm.

Specification requirements for various categories of application vary widely. They may be mainly prescriptive in terms of fiber type (or even specific brand) and amount, with the only performance requirement being compressive strength to monitor the quality of the matrix, typically when the specifier is unfamiliar with the technology of FRC or has decided to rely on the recommendations of a particular fiber manufacturer. Alternatively, they may be primarily performance-based with criteria usually defined for flexural loading that can include first-crack flexural strength to establish the stress at the onset of cracking in the matrix, ultimate flexural strength (modulus of rupture) to assess load capacity or sometimes to make a comparison with the unreinforced equivalent, toughness parameters to assess energy absorption capability, and residual strength retained after a specified amount of deflection or cracking has been sustained as described previously in section 4.3. Flexural performance parameters are most frequently based on simply supported beam tests, primarily because of historical precedents for concrete without fibers and a desire to minimize experimental complexity and testing expense. Nevertheless, a two-way slab test is more representative of the flexural loading typical of many applications, and some authorities in Europe have adopted this approach for performance evaluation of FRC.

7.1. SPECIFICATIONS FOR FIBER-REINFORCED CONCRETE

ASTM specifications for fiber-reinforced concrete are, like their counterparts for fiber-reinforced cement, in a continual state of evolution that started in 1989 with ASTM C 1116 (ASTM, 1994). It covers the manufacture of all forms of mechanically mixed fiber-reinforced concrete and dry or wet-mix shotcrete regardless of the fiber type and concrete matrix ingredients used. It is similar in principle to ASTM C 94 (ASTM, 1994) for ready-mixed concrete in that the performance parameters to be evaluated are identified without stipulating minima or grades of performance as, for example, in ASTM C 1186 (ASTM, 1995) for non-asbestos fiber-cement products.

An ACI report (ACI Committee 544, 1993) offers guidance for specifying FRC which reflects past practice specifically for steel fiber-reinforced concrete (SFRC), in which compressive strength has normally been specified for structural applications and flexural strength specified for paving applications. However, neither compressive strength (Fig. 4.12) nor first-crack flexural strength (Fig. 4.22), are substantially improved by the less than 1.5% by volume of fibers typical of most FRC's because they depend primarily on the strength of the parent matrix. Ultimate flexural strength or modulus of rupture (MOR) is usually little different from first-crack flexural strength for most FRC's because they exhibit strain-softening

(Fig. 4.14), unlike fiber-reinforced cements which have fiber contents large enough to produce significant strain-hardening with the ultimate flexural strength substantially greater than the first-crack strength or proportional elastic limit (PEL) (Fig. 4.20, 6.8, 6.15 and 6.27). Consequently, the effect of fibers in most FRC's is seen only in their load-deformation behavior after first crack as improvements in post-crack toughness and residual strength (Fig. 4.11 and 4.24). The difficulty of understanding the significance of toughness in relation to structural performance, except perhaps in the context of structures designed for energy absorption rather than simply load capacity, for example earthquake or explosive blast, has led to a preference for specifying post-crack performance in terms of residual strength at a specified end-point deflection rather than in terms of toughness expressed in units of energy to the same deflection. However, the user can be allowed to choose between the two options, as in a European specification for shotcrete (EFNARC, 1996) where performance is specified either in terms of residual strength in a beam test or in terms of energy absorbed in a two-way slab test.

7.2. STANDARD BEAM TESTS FOR EVALUATING PERFORMANCE

The various alternatives for defining post-crack performance that have been standardized, or at least widely accepted in most countries, are based mainly on flexural loading of beam specimens, with performance parameters expressed in terms of toughness or residual strength derived from the resulting load-deflection relationship. A thorough discussion of these standards and other related alternatives is necessary in order to understand the limitations and possibly misleading nature of some of the published data on flexural performance in beam tests.

In North America, ASTM C 1018 (ASTM, 1994), first adopted in 1984, provides for the determination of toughness indices which are measures of toughness relative to the toughness up to first crack, i.e. the point on the load-deflection relationship at which it first becomes nonlinear (analogous to the terms proportional elastic limit, PEL, and limit of proportionality, LOP, applied to fiber-reinforced cements). Toughness defined as the area up to a specified end-point deflection, $(n+1)\delta/2$, expressed as a multiple of the first-crack deflection, δ, is divided by toughness up to the first-crack deflection to obtain an index, I_n, which is a dimensionless number independent of the choice of units used to measure load and deflection (Fig. 7.1, left) (Johnston, 1991 and 1994–1).

As a response to the preference for post-crack performance based on strength rather than on energy ASTM C 1018 was revised in 1989 to provide for the determination of residual strength factors derived directly from toughness indices, for example $R_{10,20}$ (Fig. 7.1, left). They represent the average strength retained over a specified deflection interval expressed as a percentage of the first-crack strength. Factors of 100 and 0 correspond respectively to perfectly elastic-plastic behavior and fully brittle behavior after first crack. These R values are also determinable directly by measuring the area for any

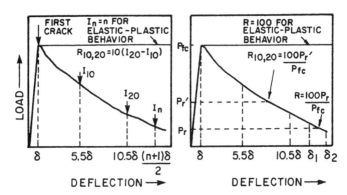

FIGURE 7.1   ASTM C1018 flexural performance parameters, in terms of toughness indices (left) and residual strength factors (right)

specified deflection interval, δ_1 to δ_2, and dividing it by the magnitude of the interval to obtain the average load ordinate for the area, P_r, which is then divided by the first-crack load, P_{fc}, to obtain the residual strength factor, again a dimensionless number independent of the choice of units used to measure load and deflection (Fig. 7.1, right). For example $R_{10,20}$ derived from toughness indices (Fig. 7.1, left) can instead be determined using the area for the deflection interval 5.5δ to 10.5δ and the average load ordinate for this area P_r (Fig. 7.1, right).

In Japan, JCI-SF4 first published in 1984 by the Japan Concrete Institute (JCI Committee, 1984) provides for the measurement of toughness, T_b, as the area up to a specified end-point deflection, δ_{tb} (Fig. 7.2), and computation of what is termed an equivalent flexural strength (JCI Committee, 1984) or flexural toughness factor (JSCE Committee, 1984). This

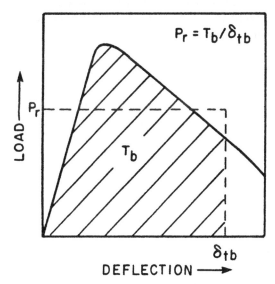

FIGURE 7.2   JCI SF-4 flexural performance parameters in terms of toughness T_b and equivalent flexural load T_b/δ_{tb}

factor is in reality an average load, P_r (Fig. 7.2), for the deflection interval from zero to 1/150 of the test span converted to an equivalent stress, $\bar{\sigma}_b$, using the elastic bending formula which is obviously not valid after first crack. It represents an averaging of the load-carrying capability before and after first crack, which is more difficult to rationalize for design purposes than a measure of the load-carrying capability over a short deflection interval immediately preceding the maximum design deflection, as is possible using an appropriate R value (Fig. 7.1, right).

Problems with both these standard beam tests that can cause them to produce erroneous or misleading results include inaccurate deflection measurement, unstable strain-softening after first crack, and a consistent objective method of defining the first crack point on the load-deflection relationship. These problems and alternative test procedures based on avoiding or addressing them are discussed in the following sections. Readers not directly concerned with the details of the requirements for proper testing of FRC beams may proceed to 7.3.

7.2.1. Deflection Measurement

Foremost of the problems encountered in flexural testing of FRC is the issue of how to properly measure beam deflection. Measurement of what is termed nominal deflection, that is deflection measured simply at the loading points (RILEM Committee 49-TFR, 1984) or at the midspan without subtracting out deflections at the supports, leads to gross overestimates of deflection, typically 2–10 times the true or net deflection (Fig. 7.3) (Johnston, 1995–1) (Chen et al., 1995). While the need to accurately measure net as opposed to nominal deflection was recognized as far back as 1985 (Johnston, 1986), and was included in the 1985 edition of ASTM C 1018, subsequent research reports often show that either net deflection measurement was not even attempted or that the results were grossly inaccurate in some cases when it was

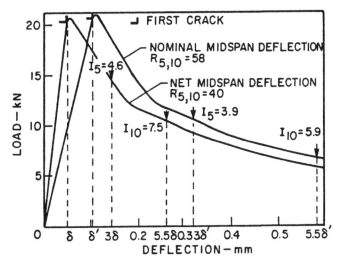

FIGURE 7.3 Load-deflection relationships and associated ASTM C1018 parameters based on nominal versus net deflection (Johnston, 1986 and 1995–1)

attempted. For gradually strain-softening behavior nominal deflection measurement can significantly lower both I and R values based on ASTM C 1018 (Fig. 7.3). However, the effect is quite variable (Fig. 7.4, Johnston, 1986 and 1995–1), and becomes negligible for cases of plastic behavior to high deflection after the initial load decrease following first crack (HE50, FP38 and FP64 fibers in Fig. 7.4). Clearly, the problem is avoided by using proper deflection-measuring equipment and verifying the accuracy of the first-crack deflection measurement as subsequently described in section 7.4.

7.2.2. Unstable Strain-Softening

The possibility of rapid unstable strain-softening immediately after first crack is of particular concern for low fiber content FRC's, typically containing 0.1% to 0.5% by volume of fibers. However, in addition to fiber content, fiber type and matrix

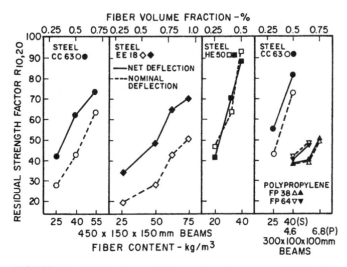

FIGURE 7.4   Residual strength factor $R_{10,20}$ based on net deflection
compared with nominal deflection (Johnston, 1995–1)

strength are also important in determining the extent of such
strain-softening (Banthia and Trottier, 1994). Under such conditions the decrease in load and increase in deflection after
first crack happen so quickly that the response rate of the
recording system may not be fast enough to reflect material
performance in the unstable region typically obtained under
conventional open-loop testing conditions. Often the instabil-
ity follows immediately after first crack (AY in Fig. 7.5), but
sometimes there is a brief interval of stable strain-softening
(AX in Fig. 7.5) that precedes the unstable interval XY
(Johnston, 1995–1). Consequently, all parameters that include
area measurements for deflection intervals between A and Y
are affected by the uncertainty introduced by unstable strain-
softening, so all ASTM C 1018 I-values and the JCI-SF4 T_b
and $\bar{\sigma}_b$ – values are affected. However, ASTM C 1018
R-values based on deflection intervals δ_1 to δ_2 that are to the
right of Y are not affected. The $R_{10,20}$ values based on net

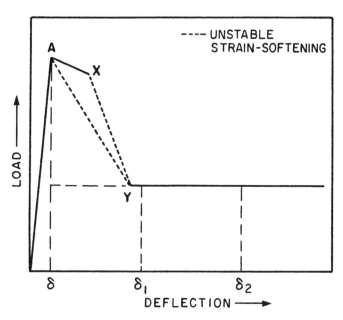

FIGURE 7.5   Effect of unstable strain-softening on the load-
              deflection relationship after first crack (Johnston,
              1995–1)

deflection measurement are examples of data obtained using
open-loop stroke-controlled testing (Johnston, 1995–1) that
clearly distinguish performance differences attributable to
fiber type, geometrical configuration and amount (Fig. 7.4).
They are unaffected by the zone of unstable strain-softening,
AY, because the response of the load-deflection recording
system is fast enough that point Y is left of the lower limit δ_1
of the deflection interval for calculating $R_{10,20}$ (Fig. 7.5).
Slower responding systems may cause the unstable zone to
extend to larger deflections thus affecting parameters such as
$R_{10,20}$ significantly. Clearly, a recording system fast enough to
ensure Y remains left of δ_1 in Fig. 7.5 is needed to avoid the
effects of unstable strain-softening on R-values.

An interlaboratory comparative testing program using five different open-loop stroke-controlled testing systems and one closed-loop system illustrates that the recording system responses can vary widely (Fig. 7.6) (Chen et al., 1995). In this case the response limit beyond which residual load values are not affected by unstable strain-softening is a deflection of about 0.62 mm. Since closed-loop testing is thought by its proponents to solve the response rate problem, a revision requiring it in ASTM C 1018 was adopted in 1997. Yet, the residual load values beyond about 0.62 mm deflection are not correlated with the response rates (Fig. 7.6), as both the highest and lowest residual loads were obtained with slow-response open-loop testing, while the four remaining essentially equal residual loads were obtained with one fast closed-loop testing system and three other open-loop systems. If the three slowest systems are excluded, it is clear that two of the relatively fast open-loop systems give residual loads about the same as the slightly faster closed-loop system at

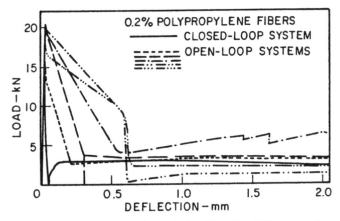

FIGURE 7.6   Load-deflection relationships for different testing systems obtained in interlaboratory comparative testing of FRC with 0.2% polypropylene fibers (Chen et al., 1995)

deflections beyond about 0.42 mm which approximates the
10.5 δ lower limit for $R_{20,30}$. Other results for low fiber
volume polypropylene FRC's confirm that residual loads
based on deflection intervals above the response limit for the
specific testing arrangement (0.75 mm in Fig. 7.7) are essentially equal for comparative open-loop and closed-loop testing
(Fig. 7.7) (Banthia and Dubey, 1996). Clearly, matching the
lower deflection limit for residual strength evaluation with the
response rate of the load-deflection recording system is more
important than whether the testing system is conventional
open-loop or the more expensive and less widely available
closed-loop type.

Another possible consequence of closed-loop testing is that
it can produce a load-deflection relationship that misrepresents
residual strength at low deflections, as, for example, in the

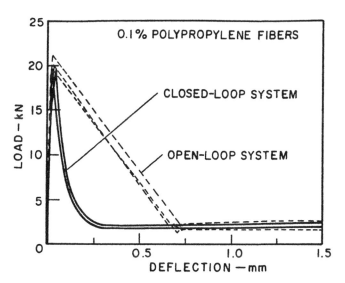

FIGURE 7.7   Comparative open-loop and closed-loop testing of
concrete with 0.1% polypropylene fibers (Banthia and
Dubey, 1996)

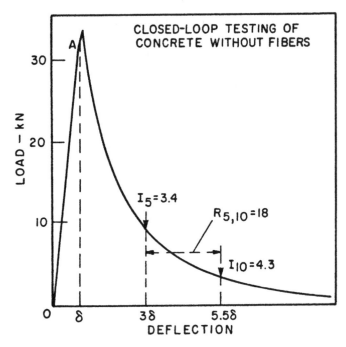

FIGURE 7.8   Load-deflection relationship for concrete without
fibers using closed-loop testing (Nemegeer and
Tatnall, 1995)

case of concrete without any fibers ($R_{5,10}$ = 18 in Fig. 7.8)
(Nemegeer and Tatnall, 1995) (Chen et al., 1995). Yet, concrete without fibers is fully brittle with zero residual strength
after cracking for practical purposes in most applications.
Thus, even closed-loop testing may misrepresent residual
strength at low deflections, just as has been illustrated for
unstable strain-softening with slow-response open-loop testing
(Fig. 7.5 and 7.6). Both examples show that the 1997 revision
requiring closed-loop testing in ASTM C 1018 is misguided
and consequently unnecessarily restrictive. Clearly, the solution is to specify only performance parameters based on
deflections larger than the response limit dictated by the

testing system, which will of course vary from one system to another (Fig. 7.6, 7.7 and 7.8). This will ensure residual strengths and R-factors that are zero for concrete without fibers, as they should be, and at realistically low values for low-fiber content FRC's (Fig. 7.7). Trials using plain concrete beams will establish the worst-case response limit for the testing system whether open-loop or closed-loop.

7.2.3. Determination of First Crack

A problem unique to ASTM C 1018 and other procedures that employ first-crack load or deflection for defining performance parameters, is the need to properly determine the first-crack point (Fig. 7.9). Despite the efforts of the standard to clarify how to do this, analysis of a magnified version of the load-deflection relationship (Fig. 7.9, left) as opposed to a more normal-scale version (Fig. 7.9, right) can substantially affect the first-crack deflection, δ, and all parameters based upon it (Nemegeer and Tatnall, 1995). The slight curvature in the

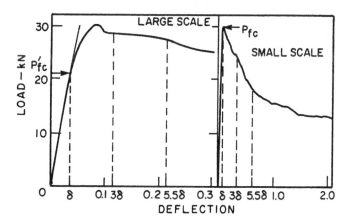

FIGURE 7.9   Magnified (left) and normal-scale (right) versions of the same load-deflection relationship (Nemegeer and Tatnall, 1995)

load-deflection relationship that is often recorded even at
normal scales can similarly affect δ. For example, in an interlaboratory comparative testing program (Chen et al., 1995)
one laboratory reported first-crack stress (mean for 24 tests on
6 mixtures with the same matrix) as 4.5 Mpa compared with
the 5.9–6.1 MPa reported by three other laboratories. Clearly,
the labs differed in their interpretation of the wording in
ASTM C 1018 that defines first crack as "the point at which
the curvature first increases *sharply* and the slope of the curve
exhibits a *definite* change" (ASTM, 1994). The difference in
interpretation is analogous to the difference between P_{fc}' and
P_{fc} in Fig. 7.9. To make the determination of first crack less
prone to subjective interpretation at least two alternatives have
been proposed.

The offset approach adopted in Belgian, Dutch, French and
German standards (Nemegeer and Tatnall, 1995) involves
drawing a prescribed offset parallel to the initial portion of the
load-deflection relationship. However, if this initial portion is
slightly curvilinear the interpretation of parallel is again subjective. The magnitude of the offset is stipulated arbitrarily by
beam size and span, for example 0.04 mm for a 100x100 mm
beam with a 300 mm span, which is approximately equal to its
first-crack deflection. First crack is determined as the point on
the curve before its intersection with the offset where there is a
decrease in load (curves B or C in Fig. 7.10), or, if such a
point does not exist, as the point of intersection of the curve
and the offset (curve A in Fig. 7.10) (Nemegeer and Tatnall,
1995).

The second alternative is a variation of the above which
bases first crack simply on the point at which a decrease in
load is first observed (Banthia and Trottier, 1995–1). This first
peak is readily determinable without difficulty of interpretation for FRC's that clearly exhibit strain-softening. Most
FRC's behave in this manner. For the few that exhibit strain-
hardening or nearly elastic-plastic behavior there may be
several peaks, and the first peak may or may not be readily

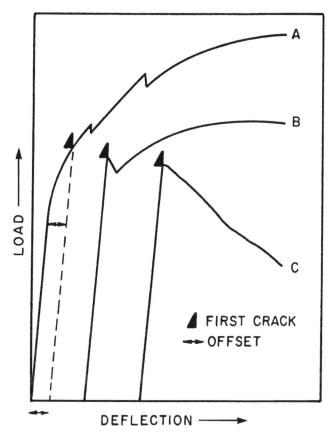

FIGURE 7.10   Offset approach for determining first crack
(Nemegeer and Tatnall, 1995)

discernible (curve A in Fig. 7.10), in which case the offset
intersection approach may be the only alternative. This behavior is more typical of fiber-reinforced cements with the higher
fiber volume fractions needed to produce strain-hardening
(Fig. 6.8, 6.15 and 6.27). Surprisingly, the issue of how to
determine the equivalent of first crack for fiber-reinforced
cements, the PEL (ASTM C 1185) and LOP (RILEM

Committee 49-TFR, 1984), evokes much less discussion and controversy than it does for fiber-reinforced concretes.

7.3. OTHER PERFORMANCE CRITERIA DERIVED FROM BEAM TESTS

To avoid the problem of determining first crack several proposals have been made that simply use arbitrarily chosen beam size-specific deflection criteria to derive residual strength parameters.

A Norwegian guide for shotcrete (Norwegian Concrete Association, 1993) stipulates a requirement for first-crack strength according to the compressive strength (C30, C40, C50 etc...) and requirements for absolute residual stresses for each performance class at specified deflections of 1 mm and 3 mm using a $450 \times 125 \times 75$ mm specimen, i.e. at 1/450 and 1/150 of the span. Class 1 is a prescription option with fiber type and amount specified. Classes 2 and 3 are performance-based with residual strengths expressed as absolute stresses (Fig. 7.11, upper). The European specification (EFNARC, 1996) for shotcrete is conceptually similar but provides for four residual strength classes and allows the end-point deflection criterion to be selected as 1, 2 or 4 mm in accordance with the deflection category (low, normal or high) appropriate to the service conditions for the application (Fig. 7.11, lower). The deflection criteria are beam size-specific, so the residual strengths derived using them are not directly applicable to other shapes and sizes of specimens.

Another proposal using arbitrarily specified beam size-specific deflection criteria for a 450x150x150 mm specimen involves determining the average residual loads for deflection intervals 0.5 to 1.5 mm and 1.5 to 3.0 mm and dividing these average loads by the first crack load to obtain R-factors that are 1/100 of the R-factors that would be obtained using the ASTM C 1018 approach for the same deflection intervals (Fig.

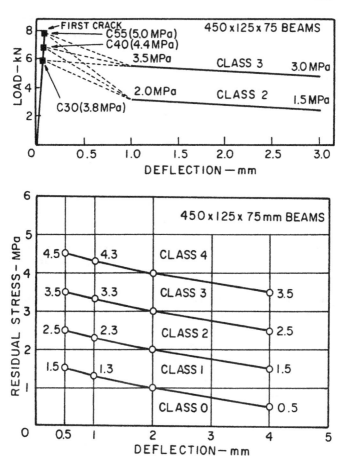

FIGURE 7.11 Performance requirements for fiber-reinforced
shotcrete in Norwegian guide (upper) and
EFNARC specification (lower) (Norwegian Concrete
Association, 1993) (EFNARC, 1996)

7.12) (Nemegeer and Tatnall, 1995). Again, neither the
deflection criteria nor the corresponding residual strength
values are readily applicable to other shapes and sizes of specimen because deflection, even under the elastic conditions that

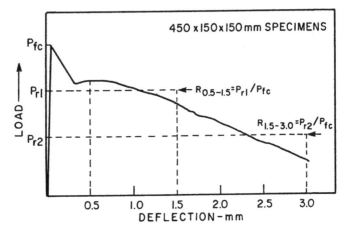

FIGURE 7.12   Residual strength factors based on averaging over a
specified deflection interval (Nemegeer and Tatnall,
1995)

prevail before cracking, is not proportional simply to span, but
involves depth as well.

Yet another proposal involves simply a modification of the
JCI-SF4 standard that determines an average flexural load and
equivalent post-crack strength (PCS) for the deflection interval
from the peak load to deflections expressed as fractions of the
test span from 1/3000 to 1/150 (Banthia and Trottier, 1995–1).
Obviously, this approach is subject to the effects of unstable
strain-softening immediately following the peak load
deflection. Using a deflection interval with a lower limit well
beyond the peak load deflection as in Fig. 7.5, 7.11 and 7.12 is
clearly preferable.

A modification of the Norwegian procedure developed for
the 300 × 100 × 100 mm beam commonly used in North
American standards advocates residual strength limits at prescribed deflections of 0.5 and 2.0 mm (1/600 and 1/150 of the
span) expressed as a percentage of the design flexural strength
and corresponding load, P_d (Fig. 7.13) (Morgan, Mindess and
Chen, 1995). The FRC under test is classified by performance
level according to the fit of the load-deflection curve to the

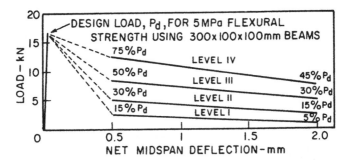

FIGURE 7.13   Residual strength factors for a 4-level template evaluation of performance (Morgan, Mindess and Chen,
1995)

prescribed limits. Only the first-crack load or equivalent stress
requirement and the deflection interval 0.5–2.0 mm are considered, so the region of possible strain-softening immediately
following first-crack is of no importance. The stipulated percentage minima for residual load shown in Fig. 7.13 are based
on extensive testing of specimens from both research and construction projects. Conformance is based on the actual flexural
strength exceeding the design value and the post-crack portion
of the load-deflection relationship exceeding the percentage
requirements between 0.5 and 2.0 mm. Again, the deflection
criteria are not directly applicable to other shapes and sizes of
specimen.

A modification to conventional open-loop testing specified
in ASTM C 1399 (adopted in 1998) is intended to address the
problem of rapid strain-softening of low-fiber content FRC's
by precracking a 300x100x100 mm beam specimen loaded
initially to 0.25–0.50 mm deflection with a 13 mm thick stiffening steel plate inserted between the specimen and the supports. After cracking, the specimen is unloaded, the stiffening
plate removed, and the specimen is reloaded in conventional
open-loop mode (Fig. 7.14) (Banthia and Dubey, 1996). The
average of the residual loads at deflections of 0.5, 0.75, 1.0
and 1.25 mm is converted to a residual stress equivalent using
the standard formula for elastic bending even though it no

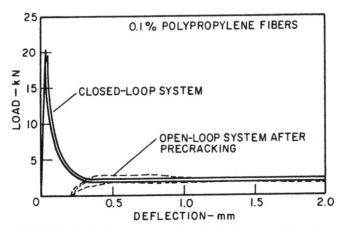

FIGURE 7.14   Residual strength determined after precracking using
a stiffening steel plate (Banthia and Dubey, 1996)

longer applies to the specimen after cracking. The results indicate that residual loads obtained using the open-loop, plate-
stiffened technique are essentially equal to those obtained for
closed-loop testing without a stiffening plate for the low-fiber
content polypropylene FRC's evaluated (Fig. 7.14). One disadvantage is that the first-crack or peak-load flexural strength
cannot be determined in the initial plate-stiffened loading
arrangement, so the residual strength cannot be related to it to
obtain the equivalent R-factor. Both the deflection criteria for
deriving residual strength and the plate stiffness (thickness)
are beam size-specific and not readily applicable to other
shapes or sizes of specimen.

7.4. REQUIREMENTS FOR PROPER
PERFORMANCE EVALUATION OF FRC IN
BEAM TESTS

Two or three requirements are essential to avoid inaccurate
and possibly misleading results when testing FRC in bending.

The first essential is accurate measurement of net mid-span deflection which is verifiable up to first crack by using the formula for elastic deflection with third-point loading

$$\delta = \frac{23\ PL^3}{1296\ EI}\left[1 + \frac{216\ d^2(1 + \mu)}{115\ L^2}\right]$$

P = peak load
E = modulus of elasticity
I = bd³/12

which for Poissons ratio $\mu = 0.2$ and span-depth ratio L/d = 3.0 reduces to (1.25) 23 PL^3/1296 EI. The 25% due to shear reduces with increase in the span-depth ratio.

The second essential is to consider performance parameters that are independent of the zone of unstable strain-softening and to verify that the response limit of the load-deflection recording system is fast enough that the performance parameters calculated from the load-deflection relationship are based on a deflection interval that does not include any of the zone of unstable strain-softening. The worst-case response limit can be established using plain concrete.

The third essential applicable only to performance parameters based on multiples of the ASTM C 1018 first-crack deflection is a procedure such as the peak load or offset approaches to more definitively and objectively determine the first-crack deflection. Performance parameters like residual strength based on arbitrarily chosen beam size-specific deflection criteria avoid the need for this requirement. Their disadvantage over performance parameters based on deflection criteria that are multiples of the first-crack deflection is that they cannot readily be applied universally to specimens of different shapes, sizes and spans, as may be desirable when comparing molded thick-section standard-sized laboratory specimens with thin-section specimens representative of overlays or shotcrete layers. In this regard, it is important to recognize that the first-crack deflection given by $\delta = (1.25)$ 23 PL^3/1296 EI for a span-depth ratio of 3.0 is equivalent to $\delta = k(\sigma/E)\ L^2/d$ where σ and E are the flexural strength and elastic modulus of the matrix (constant for a given matrix) and k is a numerical constant. Accordingly, δ is proportional to L^2/d at least up to first crack. What happens after cracking is analytically difficult to predict, but there is no rationale to support end-point deflections based simply on L.

Regardless of the deflection criteria, basing performance not simply on residual load or equivalent flexural strength, but expressing it in percentage terms relative to the first-crack load or stress makes the results more universally applicable. Universally applicable does not imply that these residual strength factors are completely independent of specimen geometry, loading and matrix variables, but they do provide a common basis of comparison that can detect the effect on performance of changes in either fiber or matrix variables.

7.5. TWO-WAY SLAB TESTS FOR EVALUATING PERFORMANCE

Performance evaluation of FRC in a two-way slab configuration more clearly represents the loading encountered in many practical applications, but is more complex and expensive than testing beams.

Two-way slab configurations have been investigated for application to shotcrete work in both Europe and Australia (EFNARC, 1996) (Bernard, 1997 and 1998). To simulate the continuity of a shotcrete layer, various support and edge restraint conditions have been evaluated to determine their influence on peak load and energy absorption capabilities, and a centre-loaded circular slab with 3-point edge support is recommended (Bernard, 1997 and 1998). The European specification for shotcrete (EFNARC, 1996) uses a 600 mm square slab simply supported along all four edges and loaded at the centre over an area 100×100 mm to produce a relationship between load and centre deflection. Areas under this relationship are used to generate an energy-deflection relationship. Performance is assessed in terms of the energy needed to produce 25 mm of deflection (Fig. 7.15), and is classified in three toughness-based categories corresponding to minimum energy levels of 500, 700 or 1000 joules.

A conceptually similar procedure for industrial floor slab-on-grade applications has been evaluated using a 3 m square slab supported on a prescribed subbase and loaded at the centre over an area of 100x100 mm (Beckett, 1990) or 120×120 mm (Falkner, Huang and Teutsch 1995). Performance is assessed in terms of load capacity rather than energy at three stages, namely the load for a first crack at the centre of the slab, the load at which radial cracks first reach the edge of the slab, and the ultimate load (Fig. 7.16) (Falkner, Huang and Teutsch, 1995).

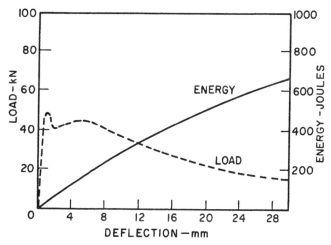

FIGURE 7.15   Load-deflection and energy-deflection relationships
for the two-way slab test (EFNARC, 1996)

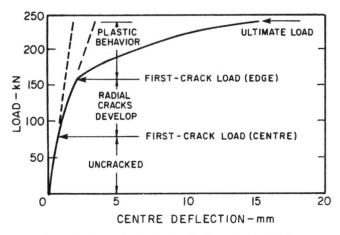

FIGURE 7.16   Stages in the load-deflection relationship for a two-
way slab (Falkner, Huang and Teutsch, 1995)

7.6. STEEL FIBER-REINFORCED CONCRETE (SFRC)

Historically, steel fibers were the type of fiber most widely
used in fiber-reinforced concrete applications, probably
because steel rates quite highly in terms of intrinsic properties
that promote superior reinforcing effectiveness (section 3.1.1),
particularly modulus of elasticity, tensile strength, ductility
and well-known compatibility with cement-based matrices.

The steel fibers of today are more performance-oriented
than the straight uniform-crossection fibers of the 1970's produced mainly either by cutting wire, much of it material
rejected by the radial tire industry, or more cheaply by cutting
thin low strength sheet material often from scrap. Nowadays,
most steel fibers produced from either wire, sheet or bulk raw
material, incorporate features to improve pullout resistance.
These include surface-roughening, surface indentation, and
crimping to a wavy rather than a straight profile over the entire

fiber length, or improving anchorage at the fiber ends by means of a hook or an enlargement that is either rectangular, spaded, or cone-shaped. Fibers can also be produced directly from a source of molten steel instead of from previously manufactured wire, sheet or other raw material. These melt-extract fibers are formed by rotation of an extraction wheel in contact with the surface of the molten steel. The surface of the wheel is designed with grooves and notches that control the length, size and shape of the fibers (Robins and Austin, 1986). Preventing oxidation and consequently inferior fiber properties is important, especially for plain carbon steels. However, the process is well suited for large scale production using wide wheels on a large surface area of melt, and it also facilitates control and modification of fiber composition because small amounts of alloying elements can readily be added. The use of small amounts of chromium and nickel can result in non-rusting fibers at a cost comparable to drawn carbon steel fibers and much below that of completely stainless fibers (Edgington, 1977). The grain size and mechanical properties of such fibers can also be altered by adjusting the melt temperature and rate of cooling. Fast cooling of the order of 100°C/s tends to produce a strong uniform fine-grained crystalline microstructure that has been particularly successful for stainless steel fibers where premature oxidation is less of a problem than for carbon steel (Hackman, 1980). The use of these relatively expensive stainless steel fibers is confined almost entirely to castable refractories. Even faster cooling rates of the order of 10^{6}°C/s have been used to produce an amorphous metallic fiber based on alloying plain carbon steel with chromium, phosphorus and silicon (de Guillebon and Sohm, 1986). These foil-like, rust-resistant fibers exist in the vitreous or glass-like state after quenching. Apparently, little is known about their long-term stability in this state where there is the possibility of slow conversion to a more stable crystalline state with a consequent change in fiber properties.

In North America, ASTM A 820 is a specification covering steel fibers for use in concrete or shotcretes. It includes minimal requirements for ultimate tensile strength (345 MPa) and ductility (bending at least 90° without breakage around a 3 mm diameter at a temperature not less than 16°C). It permits all but the weakest and most brittle of fibers to be approved. It also does little to ensure the absence of contaminants in the form of coatings to the original wire or sheet. Aluminum is a notable example of a coating that can detrimentally affect concrete by generating hydrogen gas during setting causing expansion and cracking. Zinc used in galvanizing is another example of a coating that can react detrimentally at the fiber-matrix interface.

7.6.1. Mechanical Properties

The amount of steel fibers used in concrete or shotcrete typically ranges from 0.25–1.0% by volume, 20–80 kg/m³, and the fiber-matrix interaction considerations discussed in Chapter 3 mean that the amount can rarely exceed 1.5% by volume of concrete, i.e. 118 kg/m³ (obtained by multiplying the fiber volume, 0.015 m³, by the density of steel, 7850 kg/m³). At these fiber contents improvements in strength in direct tension, compression and flexure are usually insignificant, as discussed in sections 4.1 to 4.3. The effect of fibers is mainly to improve strain capacity and consequently toughness and residual strength after matrix cracking. It is usually assessed in flexure which is the mode of loading most relevant to many applications.

7.6.1.1. Slow Flexure

The performance of SFRC under slow single-cycle flexural loading, sometimes called static loading, is most commonly evaluated using beam test data, but there is some data for two-way slab tests.

Naturally, fiber content is a factor affecting performance, as shown for the relatively short 100 mm or 150 mm square beams of span-depth ratio 3.0 typically tested in North America (Fig. 7.4, Johnston, 1995–1) and for the longer, shallower beams typically tested in Europe, for example the 150 × 100 mm specimens of span-depth ratio 7.5 used in Sweden (Fig. 7.17) (Johnston and Skarendahl, 1992). Its influence varies considerably between different fibers, and variables such as end anchorage configuration, fiber profile and aspect ratio are often more important. Despite differences between the two sets of data in terms of span-depth ratio, beam size and shape, and degree of preferential fiber alignment of long fibers confined within a small crossection, a comparison of results for various hooked-end fibers from Fig. 7.4 and 7.17 shows the expected differences attributable to fiber aspect ratio (Fig. 7.18, left). In this comparison the effect of differences

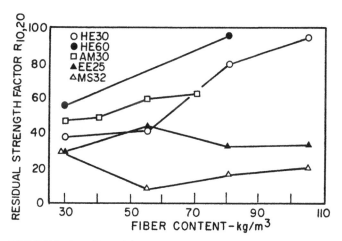

FIGURE 7.17   Effect of fiber content on residual strength factor for 25 to 60 mm long steel fibers (HE-hooked wire, AM-amorphous, EE-enlarged-end slit sheet, MS-milled shavings (Johnston and Skarendahl,

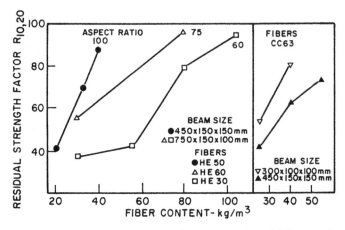

FIGURE 7.18   Effects of aspect ratio and beam size with long steel
fibers on residual strength factor (Johnston, 1995–1)
(Johnston and Skarendahl, 1992)

due to preferential fiber alignment is probably limited to the
60 mm fiber length in a 100 mm deep beam where the ratio of
minimum crossectional dimension to fiber length is 1.67,
which is less than the 3.0 minimum required by ASTM C 1018
except for specimens representing thin-section applications.
When this ratio drops from 2.38 for 63 mm crimped crescent-
shaped (CC) fibers in a 150 mm crossection to 1.58 for the
same fibers in a 100 mm crossection the measured residual
strength factor increases noticeably due to the greater degree
of preferential fiber alignment (Fig. 7.18, right).

    Fiber content has also been shown to affect performance in
two-way slab tests where both the energy absorbed to 25 mm
deflection and the residual strength retained at any specific
deflection tend to increase with increasing fiber content,
although the differences in residual strength become minimal
at high deflections (Fig. 7.19) (Clements, 1996).

    More important than fiber content is the fiber configuration
which includes characteristics such as aspect ratio, profile,
surface texture, end anchorage and even intrinsic tensile

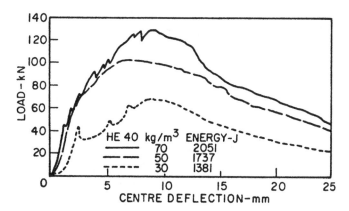

FIGURE 7.19 Effect of steel fiber content on two-way slab performance (Clements, 1996)

strength and microstructure (amorphous, fine-grained, coarse-grained etc.). In beam tests these parameters influence both the rate and degree of strain-softening to higher deflections well beyond any zone of unstable strain-softening after first crack. Fibers without end anchorage improvement tend to be associated with continual strain-softening and significant rate of decrease in residual load as deflection increases (Fig. 7.20, right) (Johnston and Skarendahl, 1992). In contrast, fibers with hooked or coned ends or a significantly crimped profile are associated with nearly constant load retention and plastic behavior to high deflection (Fig. 7.20, left) (Fig. 7.21) (Banthia and Trottier, 1995–2). Compared with end anchorage or profile improvements, changes in aspect ratio (46 to 75 in Fig. 7.21) and surface texture are probably less important.

Fiber configuration has also been shown to affect performance in two-way slab tests with hooked and crimped fibers superior to those of similar aspect ratio without substantive improvements in profile or end anchorage (Fig. 7.22) (Clements, 1996). In addition, for fibers of similar profile and end anchorage, a higher intrinsic tensile strength is beneficial, and may in part contribute to the superior performance of the

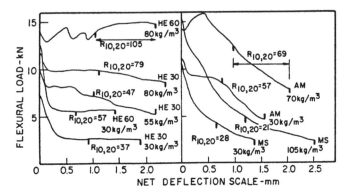

FIGURE 7.20   Post-crack performance of SFRC beams for various
                fiber configurations (as described in Fig. 7.17) and
                amounts (Johnston and Skarendahl, 1992)

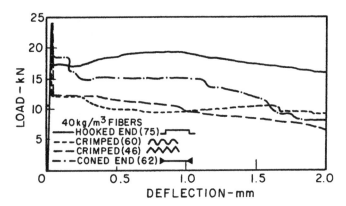

FIGURE 7.21   Post crack performance of SFRC beams for various
                fiber configurations (Banthia and Trottier, 1995–2)

hooked fibers (Fig. 7.23) (Clements, 1996). This is consistent
with the detrimental effect of heat-annealing on fiber strength
and pullout resistance mentioned in 3.2.5.

Matrix strength can also influence flexural performance, but
its influence is variable and a function not just of the matrix
alone but of the composite as a whole in terms of fiber content

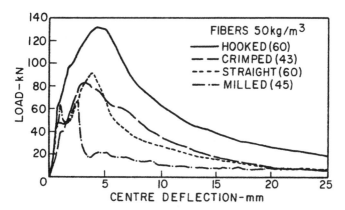

FIGURE 7.22   Effect of steel fiber configuration on performance of two-way slabs (Clements, 1996)

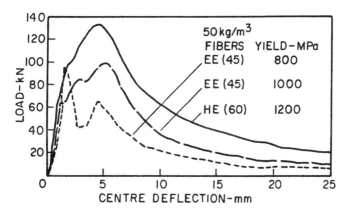

FIGURE 7.23   Effect of steel fiber tensile strength on performance of two-way slabs (Clements, 1996)

and configuration. For a moderate fiber content of 40 kg/m^3 (0.5% by volume), an increase in matrix strength can cause the residual load at deflections well beyond any zone of unstable strain-softening to decrease significantly, decrease only slightly, or increase significantly, as seen for hooked, crimped

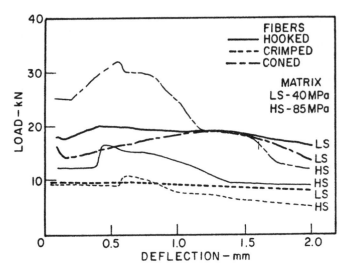

FIGURE 7.24   Effect of matrix strength on residual strength for
hooked, crimped and coned steel fibers (Banthia and
Trottier, 1995–2)

and coned fibers respectively (Fig. 7.24) (Banthia and Trottier, 1995–2), so fiber configuration is clearly important. Nevertheless, the dominant trend in results from two sources (Banthia and Trottier, 1995–2) (Balaguru, Narahari and Patel, 1992) is a decrease in residual strength factor with increase in matrix strength (compressive or first-crack flexural), although the trend becomes less severe as fiber content increases from 40 to 100 kg/m³ (Fig. 7.25).

7.6.1.2. *Flexural Impact*

Qualitatively, there is a consensus that steel fibers improve impact resistance, whether the improvement is measured simply by manually breaking samples with a hammer, using the simplistic empirical drop-weight test (ACI 544, 1988–1) where the net energy input to the specimen is not determined, or using a properly instrumented test that allows for energy

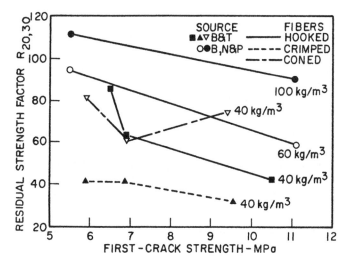

FIGURE 7.25   Effect of matrix strength on residual strength factor $R_{20,30}$ for hooked, crimped and coned steel fibers (Banthia and Trottier, 1995–2) (Balaguru, Narahari and Patel, 1992)

losses to the impacting device and the specimen support system.

Most instrumented test procedures employ the weighted pendulum (Charpy) technique (Hibbert and Hannant, 1981) (Gopalaratnam and Shah, 1986) (Banthia, Mindess and Trottier, 1996), but the drop-weight technique has also been used (Banthia, Mindess and Bentur, 1987). All vary considerably with respect to the mass and velocity of the impacting device, the maximum strain rate produced, and the specimen size. Therefore, it is not surprising that the reported quantitative improvements in impact resistance vary widely because, as has been acknowledged (Banthia, Mindess and Trottier, 1996), a significant amount of additional research is needed before the true influence of machine characteristics on the measured fracture energy values can be understood.

A conservative and simple assessment of the effect of fibers in impact loading is that increases in flexural impact energy attributable to steel fibers are at least equal to the corresponding increases in energy determined under slow flexural loading (Fig. 7.26) (Johnston, 1986) (Hibbert and Hannant, 1981). This assessment is based on 0.6 or 1.2% of hooked or crimped 50–60 mm long steel fibers in 100 mm square beams with energy computed to 10 mm of deflection for both slow flexure and impact.

An apparently conflicting report (Banthia, Mindess and Trottier, 1996) for fiber contents of 0.5% of hooked, crimped or coned 52–62 mm long fibers, tabulates impact energies

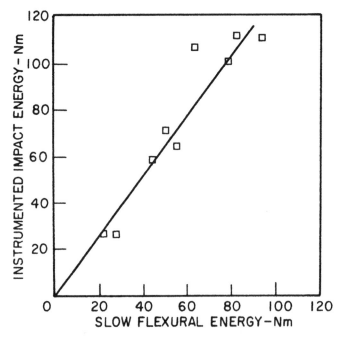

FIGURE 7.26   Relationship between impact and slow flexural energies to 10 mm deflection (Hibbert and Hannant,
1981) (Johnston, 1986)

which are 6 to 12 times the corresponding slow flexure energy. However, on closer examination it appears that impact energy is based on complete specimen fracture into two pieces, i.e. very large deflection, while slow flexure energy is based on only 2 mm of deflection which is much less than the energy needed for a very large (near-fracture) deflection. Furthermore, impact tests were performed on 75 mm square specimens while slow flexure tests were done using 100 mm square specimens, so the degree of preferential fiber alignment is greater in the impact specimens thus tending to enhance their performance relative to the slow flexure specimens. Finally, the conversion of the slow flexure energies to the smaller 75 mm square cross-section on a unit area basis may not be entirely justifiable, as acknowledged by the authors. Clearly, the slow flexure energies would more closely approach the impact values if they were compared at the same deflection and degree of fiber alignment as determined by specimen size and fiber length.

7.6.1.3. Flexural Fatigue

While there is consensus that steel fibers improve flexural fatigue performance, the improvements claimed depend on the basis of comparison. Sometimes they are stated relative to a concrete without fibers which is usually not equivalent in all respects to the matrix of the fiber-reinforced concrete, and sometimes they are related to either the first-crack or ultimate strength of the fiber-reinforced concrete (Tatro, 1987) (Ramakrishnan, Wu and Hosalli, 1989) (Johnston and Zemp, 1991). Since fatigue failure follows almost immediately after the appearance of cracking, the improvements imparted by fibers in fatigue, as distinct from any improvements they impart in static (slow flexure) loading, are probably best illustrated in terms of performance relative to the first-crack strength under static loading. However, for most SFRC's that exhibit strain-softening after cracking the difference between using first-crack strength and ultimate strength (modulus of rupture) as the basis of comparison in relationships between stress and number of load repetitions is small, and there is of course no difference for concrete without fibers (Fig. 7.27) (Johnston and Zemp, 1991).

The influence fiber content can be portrayed in two ways. Relationships that express stress as a percentage of the first-crack or ultimate strength are good for overall comparisons

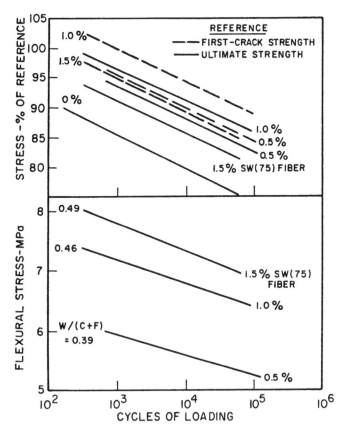

FIGURE 7.27   Effects of steel fiber content on fatigue performance (Johnston and Zemp, 1991)

between different data sets (Fig. 7.27, upper). Relationships involving actual stress are more useful for design purposes (Fig. 7.27, lower). The influence of fiber configuration is similarly presented (Fig. 7.28) (Johnston and Zemp, 1991). The behavioral trends may not be identical in each form of presentation, as, for example, with increasing fiber content (Fig. 7.27) or change in fiber configuration (Fig. 7.28).

Extended tests to 2 million cycles of nonreversing load help to explore the question of whether SFRC, unlike plain con-

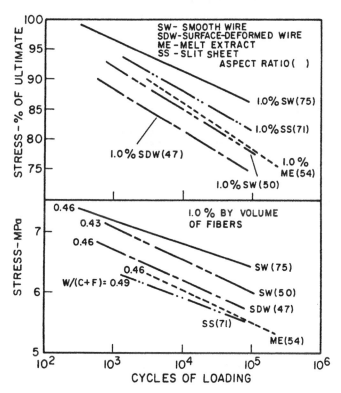

FIGURE 7.28   Effect of steel fiber configuration on fatigue performance (Johnston and Zemp, 1991)

crete, exhibits a true fatigue limit beyond which it can sustain a certain level of stress indefinitely (Tatro, 1987) (Ramakrishnan, Wu and Hosalli, 1989). Most of the data available for comparison are in the form of percentage relationships relative to the static ultimate strength, and they show that the percentage of ultimate flexural strength sustainable to 2 million cycles ranges from 50% to 85% depending on steel fiber amount and configuration (Fig. 7.29). The lower end of this range, 50–55%, is typical of concrete without fibers which has been shown not to have any semblance of a fatigue limit up to 10 million cycles of loading (Hsu, 1981). The higher end of the range, 65–85%, is characteristic of 0.5–1.0% of certain hooked and straight wire fibers. If allowance is made for dif-

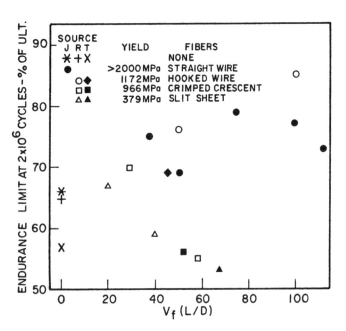

FIGURE 7.29   Influence of steel fiber strength on 2 million cycle endurance limits (Sources identified in text)

ferences in fiber content and aspect ratio by considering the 2 million cycle endurance limits in terms of the product of fiber content and aspect ratio, V_f (L/D), these higher performance fibers appear to have in common a higher yield strength than the others with which they are compared, i.e. slit sheet and crimped crescent-shaped material (Fig. 7.29). The comparison is based on actual 2 million cycle data (Tatro, 1987) (Ramahrishnan, Wu and Hosalli, 1989) and on 500,000 cycle data extrapolated to 2 million cycles (Johnston and Zemp, 1991), identified respectively as sources T, R and J in Fig. 7.29.

The apparent importance of fiber yield strength is consistent with the idea that the fatigue performance of SFRC depends to a significant extent on the fatigue performance of the steel. Accordingly, for any given level of flexural stress in the SFRC and corresponding maximum fiber stress, the higher the fiber strength the lower the fiber stress as a percentage of it, and, considering just the fatigue performance at the steel, the greater the number of cycles sustainable. Alternatively, for a constant fiber stress as a percentage of fiber strength, for example at 2 million cycles, the higher the fiber strength the greater the fiber stress sustainable for the steel and the greater the corresponding level of flexural stress sustainable in the SFRC.

The apparent importance of steel fiber strength in fatigue is consistent with its influence on fiber pullout resistance discussed previously in 3.2.5. and its influence on the energy absorption of two-way slabs under slow flexural loading (Fig. 7.23, section 7.6.1.1), so it appears that compositional raw material characteristics and manufacturing processes that improve strength are desirable for steel fiber production. This is not intended to imply that steel strength is the only important governing factor. End anchorage improvement by hooks, cones etc. is also important, as has been shown for slow flexure, especially when present in combination with high fiber strength.

7.6.2. Resistance to Cracking

Steel fibers like other fibers inhibit or modify crack development under conditions that produce tensile stress, either directly as a result of application of load, or indirectly as a result of shrinkage under restraint. For example, under slow flexural loading single or multiple cracks develop as load increases, and end-point deflection criteria for residual strength or toughness parameters can in principle be selected to match anticipated serviceability conditions in terms of maximum allowable crack width (Fig. 7.30) (Johnston, 1991). However, most attempts to relate crack development to fiber parameters such as amount, type and geometry have employed some form of restrained shrinkage test.

The restrained shrinkage test is usually based on molding a SFRC test specimen around a steel ring or core (Malmberg and Skarendahl, 1978) (Swamy and Stavrides, 1979) (Grzybowski and Shah, 1990). After normal moist curing the specimens are subjected to a prescribed drying regime producing shrinkage that is restrained by the steel ring or core. Cracks develop as drying progresses, and are monitored in terms of number, spacing and width (Fig. 7.31) (Malmberg

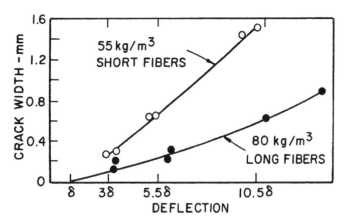

FIGURE 7.30   Crack development in SFRC specimens under slow
flexural loading (Johnston, 1991)

and Skarendahl, 1978). Free shrinkage in terms of the length change of standard prisms can be monitored concurrently, primarily to contrast the minimal effect of fibers on free shrink-

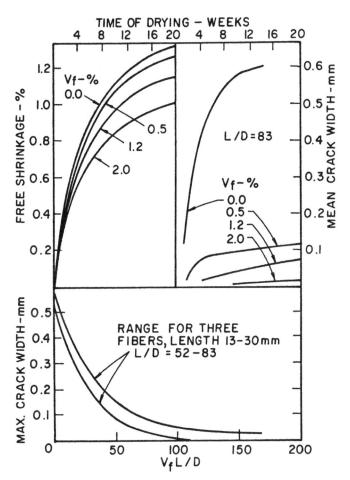

FIGURE 7.31   Effect of steel fiber amount and aspect ratio on free shrinkage and crack development under restraint for drying at 20°C and 50% R.H. after 3 days moist curing (40 × 40 mm crossection) (Malmberg and Skarendahl, 1978)

age with their much more marked effect when shrinkage is restrained (Fig. 7.31). As little as 0.25 to 0.5% of steel fibers can significantly reduce both the mean crack width (Fig. 7.31) and the maximum crack width (Fig. 7.32) (Grzybowski and Shah, 1990), with higher aspect ratio and increasing fiber content combining to enhance their effectiveness (Fig. 7.31), as in the case of flexural properties under load.

Heating of the restraining ring is an option to accelerate the cracking process by thermally expanding the ring rather than relying just on drying shrinkage, (Kovler, Sikular and Bentur, 1993), and may be quite realistic in view of the cooling and

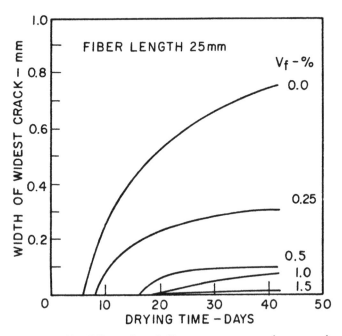

FIGURE 7.32   Effect of steel fiber content on maximum crack width for drying at 20°C and 40% R.H. after 4 days moist curing (35 mm thick × 140 mm deep rings) (Grzybowski and Shah, 1990)

drying regimes to which slabs on grade are often exposed. This procedure also helps to shorten the rather long time period of several weeks or months otherwise needed to produce significant cracking by drying shrinkage alone. Maximizing the sensitivity of the SFRC to cracking depends not only on the drying and heating (if any) regime, but on the complex relationship between the tangential tensile stress induced by the restraint mechanism, the elastic moduli of the SFRC and the restraining core, and the relative radii of the specimen and the core (Kovler, Sikular and Bentur 1993). Creep of the specimen under tensile stress is an additional stress-relieving factor that over time reduces the cracking tendency, and it too is a reality in many practical applications. Inducing the maximum possible strain in the test specimen in the shortest possible time by rapid drying of a thin test specimen restrained by a much larger and stiffer ring or core maximizes sensitivity to cracking. Naturally, these requirements conflict with other considerations such as the need to minimize the influence of aggregate maximum size and preferential alignment of long fibers in small crossections, so quantitative comparisons between different thermal or drying regimes involving different specimen geometries are not possible. Most ring specimens are 35–75 mm thick, so some degree of preferential fiber alignment in the tangential direction is unavoidable, for example with fibers up to 30 mm long in a 40 mm thick section (Fig. 7.31) or 25 mm long fibers in a 35 mm thick section (Fig. 7.32). Therefore, ring specimens can be expected to develop somewhat less severe cracking than thick-section elements on the basis of differences in fiber alignment. However, the thick sections dry and shrink more slowly allowing more time for creep to alleviate stress, which may more than compensate for any differences in cracking due to preferential fiber alignment. Predicting the resistance to cracking of thick-section units in, for example, slab-on-grade applications, from tests on thin ring specimens is therefore very difficult.

Alternatives to the ring specimen are slabs restrained at the edges (Kraai, 1985) or slabs with a centrally placed restraining ring or core (Padron and Zollo, 1990). These have been used mainly to assess the plastic shrinkage cracking tendency in specimens subjected to drying instead of normal moist curing immediately after casting, and their suitability for evaluating later-age drying shrinkage has not been established. However, a system based on one-dimensional restraint of a long prismatic SFRC specimen (500x40x40 mm) has been used to evaluate the combination of plastic shrinkage and early-age drying shrinkage that occurs within 24 hours (Banthia, Azzabi and Pigeon, 1993 and 1995). In the one-dimensional system, the restraint is provided by steel bars molded into the ends of the specimen and linked to a rigid surrounding frame. The degree of restraint is such that a severe drying environment based on exposure to air at 50°C and less than 50% relative humidity immediately after demolding at age 3 hours was found necessary to induce significant cracking, so the technique is probably unlikely to be suitable for evaluating later-age drying shrinkage cracking at normal temperatures and humidities. However, it confirms the effectiveness of steel fibers of length 3–32 mm in moderating crack development under the conditions of the test up to 24 hours after casting, which represent some combination of plastic shrinkage and early-age drying shrinkage (Fig. 7.33) (Banthia, Azzabi and Pigeon, 1993 and 1995). Again, significant longitudinal fiber alignment can be expected using fibers up to 32 mm long in a 40x40 mm crossection, and it is not possible to identify the importance of differences in fiber geometry, aspect ratio, and tensile strength individually from the data presented, except to conclude that very short (micro) fibers are probably rather less effective than the longer conventional types.

7.6.3. Durability of Steel Fibers in Concrete

Steel fibers, like conventional reinforcing steel, are completely compatible with the alkaline environment that normally pre-

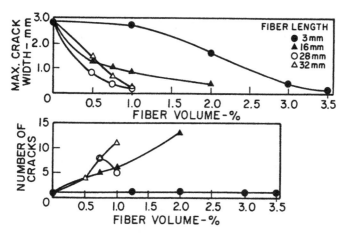

FIGURE 7.33   Effect of steel fiber content on crack development for various fiber lengths with drying at 50°C and less than 50% R.H. after 3 hours (40 × 40 mm square prisms) (Banthia, Azzabi and Pigeon, 1993 and 1995)

vails in uncracked concrete. Only when the alkali is leached out, or when the cement paste becomes carbonated, does the pH drop sufficiently for corrosion to occur in uncracked concrete. Even under conditions of marine exposure, corrosion of steel fibers has not been observed beyond the surfacial carbonated zone. The splash or tidal zones that include wetting and drying cycles, sometimes with freezing and thawing, are more critical for fiber corrosion than the fully submerged zones, just as for conventionally reinforced concrete (Schupack, 1986) (Hoff, 1987) (Mangat, 1988). The crack-inhibiting effect of fibers may also help to protect conventional reinforcement, when present, since it has been reported that corrosion of 10 mm diameter bars with 30 mm cover in a seawater splash zone was much less severe when using a SFRC matrix than when they were covered with plain concrete (Kobayashi, Hoshino and Tsuji, 1990).

Steel fibers do not significantly enhance performance under cycles of freezing and thawing (Balaguru and Ramakrishnan, 1986), although they appear to moderate the associated degree of volume increase and loss of dynamic modulus, especially for SFRC's with inadequate entrained air (Gram, Fagerlund and Skarendahl, 1978). However, fibers are not a substitute for a proper air void system obtained by using an air-entraining admixture and normal maximum limits on water-cement ratio appropriate for the exposure condition. They merely inhibit rather than prevent the propagation of cracks caused by freezing and thawing.

Any condition that leads to cracking of SFRC renders it more vulnerable to whatever corrosive agents can reach the fibers directly through the cracks. Exposure of precracked specimens in a moist marine atmosphere showed that fiber corrosion was not serious for crack widths less than 0.15 mm (Hannant and Edgington, 1975). However, more severe exposure involving circulated calcium chloride solution showed severe fiber corrosion at crack widths as low as 0.05 mm (Paul, 1976). There is probably no absolutely safe maximum crack width for steel fibers (Mangat, 1987), and corrosion can be expected to some extent in cracked concrete depending on fiber size, crack width and the severity of the exposure condition.

Since the maximum fiber stress in SFRC for fiber pullout is often initially much less than the fiber strength, steel fibers may continue to sustain their full pullout load across cracks for a considerable time before their diameter reduces to the point where the fiber stress reaches the fiber strength. It follows that fibers of large crossectional area and high tensile strength may be superior for long-term durability. However, fibers with improvements such as end anchorage, crimping or surface texturing that cause the initial fiber stress for pullout before corrosion to be close to the fiber strength may be expected to quickly change the failure mode of the SFRC from fiber pullout to fiber breakage as corrosion proceeds. When

this happens noticeable reductions in ultimate flexural strength
and significant reductions in post-crack toughness and residual
strength may be expected in accordance with the severity of
the exposure conditions and the consequent rate at which corrosion reduces the fiber diameter.

Confirmation of the effects of corrosion in cracked SFRC
has been obtained using 2% by volume of 30 × 0.5mm hooked
fibers in precracked flexural specimens subjected to wetting
and drying cycles in 3.5% sodium chloride solution (the concentration similar to seawater) at 20, 50 and 80°C, and in one
case accelerated carbonation induced by containment within a
carbon dioxide chamber for 3 weeks (Kosa and Naaman,
1990). The results demonstrate the reduction in fiber diameter
with exposure time (Fig. 7.34), along with the reductions in
strength and toughness based on an index derived using the
area under the load-deflection curve to 13 mm deflection
(Fig. 7.35). Clearly, even under quite severe corrosion condi-

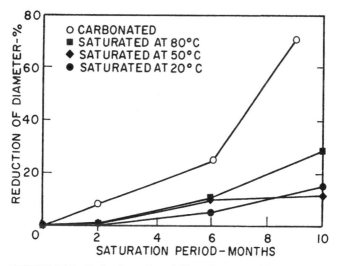

FIGURE 7.34   Reduction of steel fiber diameter with time under
various corrosion conditions (Kosa and Naaman,

tions the retention of ultimate strength in cracked SFRC can be considerable, better than 85% for wetting and drying cycles in salt solution for 10 months and better than 75% for specimens subjected to carbonation and cycled for 9 months. Corresponding toughness retention is also substantial at better than 60% for wetting and drying in salt solution and 40% for wetting and drying preceded by carbonation (Fig. 7.35) (Kosa and Naaman, 1990).

Obviously, the retention levels can be expected to vary for fiber amounts and geometries different from the 2% (160

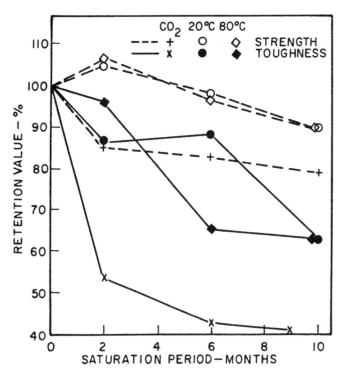

FIGURE 7.35   Loss flexural strength and toughness under various
corrosion conditions (Kosa and Naaman, 1990)

kg/m^3) of 30 x 0.5 mm hooked fibers examined above. For similarly tested slurry infiltrated (SIFCON) specimens with 14% by volume of 30x0.5mm hooked fibers strength retention is reported as at least 90% and toughness retention as at least 70% (Kosa, Naaman and Hansen, 1991). Accordingly, it appears that retention of strength and toughness may be less at the lower fiber contents below 1.0% (80 kg/m^3) typical in many applications. Consequently, for very severe corrosion conditions where cracking of SFRC is anticipated, the use of alloyed steel fibers (de Guillebon and Sohm, 1986) (Edgington, 1977) may be necessary to fully address the possibility of significant losses of strength and toughness by fiber corrosion.

7.6.4. Steel Fiber-Reinforced Concrete (SFRC) Applications

The early applications up to 1975 included slab-on-grade and overlay construction for airfields, highways, bridges and industrial floors, repairs to cavitation or erosion-damaged zones in dam stilling basins and spillways, shotcrete applications in mining, tunnelling and slope stabilization, and production of a variety of precast concrete products such as pipe, utility poles and dolosses for wave protection (Lankard, 1975). Pavement overlay and shotcrete applications comprised a large proportion of the total applications up to early 1980's (Johnston, 1982), many of which are included in a listing of 112 projects worldwide compiled by ACI Committee 544 and made available by ACI (Henager, 1981–2).

A 1996 ACI state-of-the-art report (ACI 544, 1996) identifies some of the more recent applications of SFRC in the form of cast-in-place construction, shotcrete, precast elements, slurry infiltrated fiber composites and refractories (using stainless steel fibers). The applications are numerous and diverse, and only a few selected examples typical of each category will be discussed.

7.6.4.1. Airport Applications

Airport pavement and overlay applications up to the mid 1980's are described in terms of size, thickness, mixture proportions and performance (Hoff, 1986) (Schrader, 1986). Generally, the objective was to achieve performance equivalent to plain or conventionally reinforced concrete with reduced thickness and increased joint spacings. A design procedure was developed from the results of field trials performed on SFRC pavements and overlays in the United States using wheel loadings simulative of large military and civil aircraft (Parker, 1974) (Parker and Rice, 1977). It is based on the U.S. Corps of Engineers concept of a standard thickness of plain concrete, defined as the thickness required for 5000 loadings by the particular aircraft type under consideration, and determined by limiting the maximum tensile stress in the slab acting parallel to the edge to 0.75 times the concrete flexural strength (to allow for load transfer) divided by a design factor of 1.3. The thickness of SFRC for 5000 or any other number of loadings is derived from this standard thickness using the function relating thickness ratio and number of loadings for the upper range of the field data obtained in controlled accelerated traffic tests (Fig. 7.36, upper) (Parker, 1974).

Most of the airport applications have been overlays, although the original field tests used to develop the design criteria involved both slab-on-grade and overlay. The two projects undertaken at Las Vegas airport in 1976 and 1979 are of interest not just because of their large size, but because the exposure condition is particularly severe with respect to extremes of temperature and humidity. The first was a 152 mm thick overlay of area 52,700 m² placed on existing asphaltic pavement without special treatment. The mixture contained 356 kg/m³ of cement and 148 kg/m³ of fly ash with 95 kg/m³ of 25 × 0.25 × 0.56 mm straight uniform slit sheet fibers. Slipform paving equipment was used in 7.6 m lanes with sawn transverse joints at 15 m intervals. The second was 178 mm

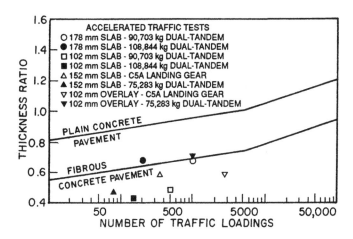

FIGURE 7.36   Performance criteria for fibrous and plain concrete pavements based on simulated aircraft loading (upper) (Parker, 1974) and slipform paving using SFRC at Las Vegas airport (lower) (Courtesy of Bekaert Corporation)

thick layer of area 91,000 m^2 placed unbonded over 50 mm of new asphaltic concrete on 300 mm aggregate base. The mixture proportions and joint spacings were essentially the same as before except that the fiber content and geometry were respectively 50 kg/m^3 and 50 × 0.5 mm hooked wire (Fig. 7.36, lower). Flexural strengths in both cases were in excess of 7 MPa. About 10% of the slabs in the 1979 project were reported as having corner breaks after 3–4 years (Packard and Ray, 1984). In the older 1976 project somewhat more than 10% had cracked, in some cases with faulting and spalling. About half the sawn transverse joints did not function properly despite sawing to a depth of 76 mm, so this caused excessively wide opening of the joints that did function along with consequent joint sealant failures. In 1980, a 70,000 m^2 overlay 178 mm thick separated from existing concrete by an asphaltic debonding layer was placed at Denver's Stapelton airport, again using 50 kg/m^3 of 50 × 0.5 mm hooked wire fibers in a matrix containing 315 kg/m^3 of cement with 150 kg/m^3 of fly ash. Joints were sawn at 12 m intervals in the 7.6 m lanes, and subsequently some curling-related corner cracks developed under load but remained tight. Slab on granular base construction was employed with SFRC 203 mm thick placed over an area of 20,200 m^2 at Reno airport in 1980. Only a few centreline longitudinal cracks were reported after 2 years (Packard and Ray, 1984). Similar slab on cement-treated base construction was adopted for a further 32,500 m^2 of Denver's 1980 rehabilitation of Stapelton airport.

In contrast to their essentially unbonded counterparts on asphaltic concrete, overlays fully bonded to existing concrete pavement with a mortar or cement paste slurry have also been constructed, for example the 29,200 m^2 of 102 mm thick overlay with 120 kg/m^3 of 25 × 0.25 × 0.56 mm slit sheet fibers at Reno airport placed in 1975. Jointing corresponded to the 3.8 × 6.1 m pattern in the original pavement. The bond apparently failed before it developed properly, probably because of premature curling due to the large day-night temperature differential following construction. Subsequently, curling at the corners led to spalls and cracks that were reported as requiring maintenance (Packard and Ray, 1984).

Overlays 127 mm thick to existing concrete pavement topped with a thin 38 mm asphaltic levelling course to produce an effectively unbonded condition were placed at military airports in Fallon, Nevada and Norfolk, Virginia in the late 1970's and early 1980's. The transverse joint spacings ranged from 12 m at Fallon to 7.6 m at Norfolk. The fibers used were either 95 kg/m^3 of 25 $\times$ 0.25 $\times$ 0.56 mm slit sheet or 50 kg/m^3 of 50 $\times$ 0.5 mm hooked wire (Schrader, 1986). Performance was reported as good to excellent (Packard and Ray, 1984), but subsequently curling-related corner cracking and some mid-slab cracking between transverse joints has been reported (Wu and Jones, 1987). Also, the problem of fibers at poorly finished surfaces possibly separating and entering jet engines became a major concern, and in the later projects led to the use of a special grid roller to push the fibers into the surface (Wu and Jones, 1987).

The three major concerns with SFRC in airfield pavement are exposed or loose fibers at the wearing surface, corner and edge curling and consequent cracking or spalling, and load transfer and jointing practices (Schrader, 1986). Actual evidence, as opposed to concern (Wu and Jones, 1987), of injury to personnel being cut or pricked by surface fibers, tire punctures, or damage associated with ingestion of loose fibers into jet engines has been almost impossible to confirm (Schrader, 1986). Proper finishing to depress fibers into the surface followed by post-construction clean-up seems all that is necessary for most applications. Sweepers with large magnets were used to improve clean-up at the Denver and Las Vegas projects. However, the issue remains a major concern for military aircraft with jet engine intakes much closer to the pavement than on large civil aircraft, and a variety of cleaning, grinding and coating techniques have been explored, none of which were considered fully effective (Wu and Jones, 1987).

Corner and edge curling effects may arise not only from severe changes in moisture content and temperature due to weather, but also from early-age thermal effects due to heat generated by cement hydration causing the top portion of a slab exposed to the atmosphere to cool faster after setting than the bottom and central portions where heat of hydration is less easily dissipated (Schrader, 1986). Observations of curl approaching 13 mm for SFRC less than 1 day old at Salt Lake City airport support the probable significance of the internal heat factor even for 203 mm thick slabs (Schrader, 1986). Many SFRC airport projects have employed cement contents of 360 kg/m^3 or more, often with 120–150 kg/m^3 of fly ash in addition, and as much as 483 kg/m^3 of cement without fly ash in the second Reno project. Obviously, mixtures with high cement content are particularly prone to heat-related curling and excessive drying shrinkage problems. Lower cement contents combined with inclusion of fly ash and the use of aggregate of at least 20 mm or greater maximum size, provided fiber length exceeds the aggregate maximum size, all help to reduce early-age curling tendency due to heat generation and later-age curling due to drying shrinkage. Later-age curling may also be addressed by increasing SFRC thickness in overlays, or by compositing the more expensive SFRC on top with a cheaper underlying concrete such as lean concrete or roller-compacted concrete. Based on corner and mid-slab cracking observed at Fallon and Norfolk, the original design criteria (Fig. 7.36, upper) that permitted thickness reduction according to number of load applications has been amended to permit no reduction over the thickness design for unreinforced concrete when using SFRC for military airport overlays (Wu and Jones, 1987).

Related to the curling problem is the omission in many of the early SFRC pavements of load transfer devices such as keyways and dowels, and the tendency to maximize longitudinal and transverse joint spacings. While there is not complete agreement on optimum joint design and spacing, experience suggests 8–12 m transverse joint spacing in placements 8 m wide or less (Schrader, 1986) (Packard and Ray, 1984), preferably using SFRC with a lower shrinkage potential as previously described because of some centreline longitudinal cracking observed in 8 m wide placements. Dowelled joints at 30 m intervals were used in 230 mm thick replacement taxiway construction at New York airports in 1992 without intermediate sawn joints. Partial-depth sawn joints must be deep enough to ensure cracking in the joint rather than adjacent to it as in the Las Vegas project in which joints were cut to 43% of the 178 mm depth. Properly designed keyways or properly aligned and functioning dowels may also be desireable (Schrader, 1986) (Packard and Ray, 1984).

Overlays introduce the question of what is the optimum bond condition with the underlying pavement. The use of asphaltic levelling course or double polyethylene sheeting has been successful in many relatively thick (125–203 mm) overlays to existing asphalt on granular base in airport applications. In these circumstances the SFRC acts more like a slab on a strong base than as a true overlay integrated with the underlying pavement. For overlays to existing concrete pavement with an asphaltic levelling course, experience indicates that debonding worked well for thick overlays at Fallon and Norfolk where joint spacings in the SFRC were much larger than in the underlying concrete. For thinner overlays full bonding is desirable to resist curling, and joints in the overlay should be matched as far as possible to the existing pavement as in the 102 mm overlay at Reno. While fully bonded overlays are less prone to curling they are more prone to cracking caused by shrinkage or thermal change, and unless the joints are matched with the joints or working cracks in the underlying pavement reflective cracking is also likely (Hoff, 1986) (Schrader, 1986). Various procedures intended to achieve full bonding are described in the following sections.

7.6.4.2. Highway and Bridge Deck Applications

The potential for thickness reduction in highway pavements using SFRC is less than in airport pavements with their heavier loadings and greater thicknesses, so SFRC tends to be economically attractive mainly for pavements carrying large numbers of heavy vehicles. Pavements subjected to large numbers of city buses which are notorious for causing rutting on major urban streets without significant truck traffic are a typical example. A 1976 test on 7th Avenue in Calgary with slabs 7.6 m long and 7.0 m wide showed that after 32 months in service equivalent performance in terms of cracking was achieved by SFRC with 45 kg/m^3 of fibers at 60–75% of the thickness needed for plain concrete or conventionally rein-forced concrete (Fig. 7.37) (Johnston, 1984–1). This was con-sistent with an earlier 1973 test with smaller 3.65 m long by 3.35 m wide slabs that after 4 years in service indicated thick-ness reductions for equivalent performance as 65–75% and 50–60% for fiber contents of 40 kg/m^3 and 80 kg/m^3 respec-tively (Johnston, 1984–1). One observation consistent with the behavior reported for airport SFRC placements (Packard and Ray, 1984) was the tendency for longitudinal centreline crack-ing in placements 7–8 m wide.

Similar full-depth comparisons of SFRC with plain concrete were made using test pavements constructed in 1988 and 1990 in Rapid City, South Dakota (Ramakrishnan, 1995). The 1988 Haines Avenue project compared 127 mm and 152 mm thick SFRC using 39 kg/m^3 of 60x0.8 mm hooked wire fibers with 191 mm thick plain concrete in slabs with longitudinal and transverse joints at 3.7 m and 4.6 m respectively. The 1990 Sheridan Lake Road project used the same concrete mixture proportions with 39 kg/m^3 of 50 mm long crimped crescent-shaped fibers of aspect ratio 40–65. Again the comparison was with 191 mm plain concrete, presumably with the same SFRC thicknesses and jointing details as before since they are not reported. Very dry hot conditions prevailed during the second day of construction of the Haines Avenue project with a day-to-night temperature differential of 30°C. This caused about 30 cracks in the 734 m of plain concrete and one in the 50 m of SFRC placed that day. No cracks were evident in either

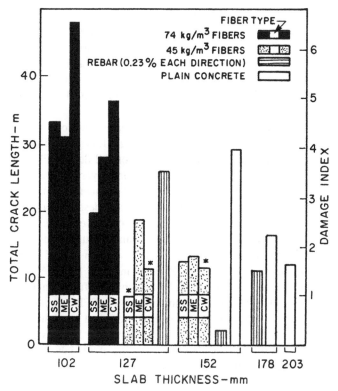

FIGURE 7.37   Performance of 7th Avenue Calgary test slabs after 32 months (SS-slit shelt, ME-melt extract, CW-cut wire) (Johnston, 1984–1)

plain concrete or SFRC placed earlier in more favorable weather conditions. Subsequent inspections revealed some new cracks parallel to and near the sawn transverse joints in the plain concrete placed under hot dry conditions. The Sheridan Lake Road was reported free of cracks in both plain and SFRC sections after 30 months. It was concluded that the SFRC pavements achieved performance equivalent to the plain concrete at 67% and 80% of the thickness (Ramakrishnan,

1995) which is consistent with the Calgary 7th Avenue data
for about the same (39–45 kg/m^3) fiber content (Johnston,
1984–1). The concrete mixture proportions are also similar
with cement and fly ash contents of 312 kg/m^3 and 113 kg/m^3
in South Dakota, 326 kg/m^3 and 89 kg/m^3 respectively in
Calgary in 1976, and 297 kg/m^3 of cement with 119 kg/m^3 of
fly ash in Calgary in 1973.

   Another approach to slab-on-grade pavement construction
with SFRC that may offer cost savings, less possibility of
curling, and smaller deflections than for full-depth SFRC, is
compositing the SFRC with an underlying layer of lower-cost
concrete such as lean concrete. A 1977 econocrete project in
Florida demonstrated the success of what is described as fully
bonded SFRC 52 mm thick on 229 mm of lean concrete, compared with the rapid delamination and failure leading to
replacement of unbonded construction with 76 mm of SFRC
on 203 mm of lean concrete (Packard and Ray, 1984). A 1977
project on 6th Avenue in Calgary achieved full composite
action by placement of a 51 mm thickness of SFRC monolithi-
cally on 279 mm of lean concrete before it had set (Johnston,
1984–1). After 7 years of service under about 500 buses per
day it was still serviceable with only nine of 33 slabs exhibiting damage amounting to one transverse crack. Fiber contents
ranged from 74 kg/m^3 of 25 mm long uniform wire, slitsheet,
or melt extract (Calgary project) to 60 kg/m^3 of 50x0.5 mm
hooked wire (Florida project). The Florida project used
350 kg/m^3 of cement without fly ash in the SFRC layer and the
Calgary project 326 kg/m^3 of cement with 148 kg/m^3 of fly
ash. The lean concrete in the Calgary project was not air-
entrained, and was made using 107 kg/m^3 of cement with
71 kg/m^3 of fly ash (Johnston, 1984–1).

   For SFRC overlays on highways the notion in many early
trials that 50–100 mm thick SFRC layers could effectively
span joints and working cracks in the existing pavement led to
unsatisfactory performance (Packard and Ray, 1984). The
bond condition was also variable with many sections placed

directly on existing concrete and classed as partially bonded, even though they subsequently became unbonded in many cases. The early 1970's projects at Eight Mile Road in Detroit, Greene County, Iowa, and the U.K. M10 Motorway are examples of such experience, demonstrating that full bonding of thin overlays and matching of joints in SFRC to those in existing pavement are essential to ensure satisfactory performance (Hoff, 1986). The rapid disintegration and failure of unbonded construction in the Florida econocrete project described above also confirmed the importance of full bonding of thin SFRC wearing courses. An example of an apparently successful fully bonded overlay is the 1983 repair on Interstate 610 near Houston, Texas in which SFRC with 50 kg/m^3 of 60x0.8 mm hooked wire fibers was placed on severely cracked continuously reinforced concrete pavement that had been in service for 16 years. Sections of SFRC of thicknesses 50 and 75 mm were compared with overlays of the same thicknesses reinforced with welded wire mesh, all bonded to the underlying pavement by scarification to about 6 mm, sandblasting, airblasting and coating with cement grout (Bagate, McCullough and Fowler, 1985). After 6 months under traffic the sections with fibers were reported as most effective in reducing cracking.

The importance of bonding was also examined to some extent in a 1986 evaluation of 18 sections of overlay placed on scarified or sandblasted concrete pavement on a heavily trafficked autoroute in Quebec (Chanvillard, Aitcin and Lupien, 1989). Nails driven into the existing pavement and protruding halfway through the depth of the new overlay were used to produce a mechanical bond. All the 75 mm thick SFRC sections (fiber content 24 or 34 kg/m^3) remained well bonded with or without nails, while the 100 mm thick sections without fibers debonded and cracked severely after 8 months whether nailed or not. The results clearly demonstrated the crack-reducing, performance-enhancing effect of the fibers, particularly at 34 kg/m^3, but were inconclusive regarding the possible benefit of the nails for improving bond.

7.6.4.3. Recent Trends in Pavements and Overlay Applications

Recent applications reflect much of the experience described above for airport and highway pavements or overlays.

In Japan, a section of Tokyo airport pavement was designed and constructed with a 15% reduction in thickness attributed to SFRC, and the satisfactory performance of SFRC in a fully bonded bridge deck overlay after 15 years under heavy traffic has captured attention for repair of roads and bridges (Sakai, Ochi and Kitoh, 1995).

In Australia, applications for SFRC include light duty pavement in car parks or tennis courts, typically with a specified flexural strength of 5.0 to 5.5 MPa and 50–62 kg/m^3 of fibers, and medium duty pavements with strength 5.5 to 6.0 MPa and 75 kg/m^3 of fibers constructed using 125–200 mm of SFRC on up to 150 mm of lean concrete (McGregor, 1995). The Australian national pavement design manual (Austroads, 1992) specifies a flexural strength of 5.5 MPa and a minimum thickness of 125 mm for SFRC compared with 150 mm for other concrete. Medium duty SFRC pavement is also used widely in roundabouts (known as traffic circles in North America) at major intersections (New South Wales, Roads and Traffic Authority, 1996). This application involves higher stresses from heavy commercial vehicles than normal straight pavement because vehicle overturning moments increase outer wheel loads in accordance with vehicle speed and turning radius to as much as 80% of total axle load. Roundabout geometry also leads to a complex joint pattern with many irregularly shaped slabs not readily amenable to dowelling and many acutely angled corners that are particularly prone to break-off. Recommended joint layouts vary with turning radius and approach designs (Fig. 7.38, upper) (NSW Roads and Traffic Authority, 1996). Field experience with these joint layouts using a minimum of 160 mm of SFRC on a lean concrete subbase shows that longer slabs and sharper corner angles are possible without compromising performance (6 m

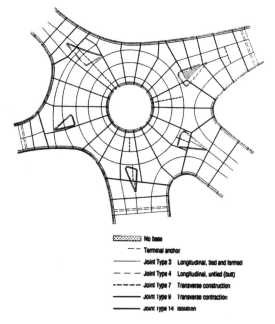

Joint Type 3 Longitudinal, tied and formed
Joint Type 4 Longitudinal, untied (butt)
Joint Type 7 Transverse construction
Joint Type 9 Transverse contraction
Joint Type 14 Isolation

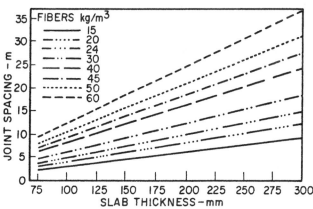

FIGURE 7.38   Joint layout for highway roundabouts (upper) (NSW
Roads and Traffic Authority, 1996) and recom-
mended joint spacing for industrial floors (lower)
(Courtesy of Novocon International Ltd.)

maximum and 70° minimum compared with 4.2 m and 85° for plain or mesh-reinforced construction), thus improving slab layout (undowelled) and construction time by reducing the number of joints.

In Canada, overlaying existing asphalt with SFRC, commonly termed white-topping, has been undertaken in two projects in Calgary, using about 34 kg/m³ of 50 mm long crimped fiber in a 100 mm overlay with 450 kg/m³ of cement. Transverse joints were sawn at various intervals from 5 m to 33 m (Johnston, 1995–2). Both overlays have served the intended purpose of combating recurrent rutting of the asphaltic pavement by city buses after 7 to 8 years in service, but transverse cracks have occurred in the longer sections, indicating that a joint spacing of 5–6 m is optimal. However, fully bonded bridge deck overlays are the largest paving application of SFRC in Canada. To achieve bond the existing concrete is sandblasted, soaked with water for at least 3 hours, brought to a saturated surface-dry state by removing excess water with compressed air, and coated with a mortar slurry immediately followed by placement of the overlay concrete. Between 1984 and 1988 about 26 overlays totaling 11,400 m² were built by the province of Alberta. (Johnston and Carter, 1989). Since 1988 a provincial highway department specification has evolved that utilizes concrete with a minimum of 350 kg/m³ of cement, 7.5% silica fume and 60 kg/m³ of fibers. It requires rigorous attention to surface preparation, slurry application, and limiting concrete placement temperature (Table 7.1) (Johnston, 1995–2). Prebagged material is used for locations remote from concrete plants. Performance has in most cases been very satisfactory. Increased deck strength, reduced dead load stresses, rapid installation with minimum traffic disruption, cost-effectiveness on a unit area basis compared with other alternatives, greater resistance to disintegration and pothole development if debonding and delamination occur, and longer service life are the advantages perceived by Alberta Transportation, which has

TABLE 7.1   Alberta Transportation bridge deck overlay
specification development for SFRC, 1988–1994

Year of operation	1988	1994
Coarse Aggregate Size	20 mm	20 mm
Minimum Cement Content	None	350 kg/m^3
Silica Fume by Weight of Cement	5%	7.5%
Fiber Content	60 kg/m^3	60 kg/m^3
Fiber Length	60 mm	50 mm
Maximum w/c+s Ratio	None	0.38
Maximum Slump	50 mm	60 mm
Superplasticizer Permitted	NO	YES
Air Content	6–8%	5.5–7.5%
Maximum Concrete Temperature	25°C	18°C
Slump Retention After 45 Minutes	None	50%
Min. Comp. Strength at 28 Days	35 MPa	35 MPa
Max. Air Void Spacing Factor	None	0.23 mm
Sand/Cement/Silica Fume for Bonding Slurry	50/50/0	50/46.25/3.75
Maximum Discharge Period	1.5 hr	1.0 hr

used about 80,000 m^2 of SFRC in about 107 bridges up to
1994 (Johnston, 1995-2).

In the United States, recent use of SFRC in bridge deck
overlays is exemplified by 14 demonstration trials (7 with
fibers and 7 without fibers) involving 70 mm thick fully
bonded construction using 60 kg/m^3 of fibers in a superplasticized matrix with either 490 kg/m^3 of cement or 415 kg/m^3 of
cement with 10% silica fume (Baun, 1993). Performance has
been excellent in the first year under an average truck loading
of 4000 vehicles per day, with no cracking in the section with
490 kg/m^3 of cement and only slight hairline cracking in the
section with 415 kg/m^3 of cement and 10% silica fume. The

SFRC is reported to be cost effective compared to alternatives such as latex-modified overlays (Krstulovic-Opara et al., 1995). Following the successful 1983 bonded highway overlay trial on Interstate 610 near Houston, Texas, an area of about 130,000 m^2 of another major route in the area called Beltway 8 was overlayed in 1996 with 50 mm of fully bonded SFRC using 46 kg/m^3 of 60x0.8 mm hooked wire fibers. Similar work is pending or in progress on highways in Louisiana and Missouri.

7.6.4.4. Industrial Floors and Overlays

The fluctuations and extremes of temperature and humidity which can cause later-age curling and drying shrinkage are usually less severe for floors that become enclosed and protected within buildings than for externally exposed applications such as airport or highway pavements. The practice of using SFRC in floors to replace the traditional welded wire mesh first became quite widespread in Europe with over 1,000,000 m^2 reported up to 1985 (Vandenberghe and Nemegeer, 1985). Fiber contents are typically 20–50 kg/m^3 (Bekaert, 1995) in Europe and North America. Properly distributed fibers reinforce both the top and bottom portions of the slab where the maximum flexural tensile stresses can occur under edge or mid-width loading, and avoid the problem of locating steel mesh reinforcement in the optimum position for these loadings and maintaining it in position during construction. Naturally, mesh placed only at the mid-depth of the slab has no reinforcing effect with respect to either loading condition. Reductions in slab thickness and increases in joint spacing are other perceived advantages over the traditional mesh-reinforced form of construction.

Floor design manuals were first developed mainly by the fiber manufacturers. Initially, they were based on allowable flexural stresses due to the design loading not exceeding a certain fraction of the ultimate flexural strength or modulus of rupture, determined according to the number of loadings, the subgrade quality and whether the exposure is interior or exterior (Bekaert, 1984) (BHP Reinforcing Products, 1994). Subsequently, this approach has been modified so that the stresses due to the design loading calculated by conventional Westergaard or similar analysis do not exceed the residual strength determined on the basis of $R_{10,30}$ or $R_{10,50}$ measured on a $450 \times 150 \times 150$ mm beam according to ASTM C 1018 (Fig. 7.1) or $\bar{\sigma}_b$ according to JCI SF-4 (Fig. 7.2) (Moens and Nemegeer, 1991) (Bekaert, 1995). For example, the allowable design stress based on $R_{10,50}$ measured on a 450x150x150 mm beam tested according to ASTM C 1018 is obtained by multiplying the ultimate beam flexural strength (almost identical to the first-crack strength) by $R_{10,50}/100$ (Bekaert, 1995). A U.K design guide for industrial floors (Concrete Society, 1994) provides for the calculation of ultimate moment of resistance using the JCI SF-4 equivalent flexural strength factor to account for the contribution of the steel fibers. These approaches are consistent with the recognition that, while the Westergaard analysis can be used to predict the cracking load for a SFRC slab, it underestimates the ultimate load considerably, and that an analysis that accounts for the plastic postcracking behavior of SFRC is more appropriate (Falkner, Huang and Teutsch, 1995) (Tatnall and Kuitenbrouwer, 1992) (Petersson, 1992).

Successful use of SFRC in floor slab also requires alteration to joint design and configuration. Sawcuts should be deep enough to ensure cracking in the joint as intended, at least 33% and possibly up to 50% of the slab thickness. Recommended joint spacings vary widely with higher values permitted for interior applications. One manufacturer recommends 15 m and 10 m joint spacings for interior and exterior applications respectively for minimally restrained slabs with a further reduction to 6 m for heavy standing loads which tend to increase frictional restraint (BHP Reinforcing Products, 1994). Another manufacturer provides a chart relating joint spacing to fiber content and slab thickness for slabs with minimal restraint placed on a slippery subgrade or on perforated polyethylene sheeting, and recommends reducing the spacing by up to 20% for slabs placed on rough granular material (Fig. 7.38, lower). This permits the spacing to vary from as little as 3 m to as much as 35 m depending on fiber content and slab thickness. Another recommends that the maximum slab size be limited in area and kept as close to square as possible, with areas corresponding to spacings of 12 m interior and 8.5 m exterior for 30 kg/m^3 of fibers and 8.5 m interior and 6 m exterior for 20 kg/m^3 of fibers (Bekaert, 1984).

Use of SFRC in large industrial floors and other slab-on-grade applications became widespread in the late 1980's and 1990's. The 7400 m^2 Honda automobile plant floor constructed in Alliston, Ontario is typical, comprising a 152 mm slab on grade reinforced with 20 kg/m^3 of 60 mm long deformed fibers (Hubler, 1986). The 140,000 m^2 floor constructed in 1990 for the Chrysler Jefferson automobile assembly plant in Detroit, Michigan is another example described in more detail (Robinson, Colasanti and Boyd, 1991). The SFRC mixture was made with 30 kg/m^3 of 60 × 0.8 mm hooked wire fibers using 366 kg/m^3 of cement with a superplasticizing admixture. The paint shop floor was 127 mm thick, accepted as equivalent to 152 mm of reinforced concrete, and the assembly shop floor was 165 mm thick, accepted as equivalent to 203 mm of reinforced concrete. Joint spacing ranged from 6.1 to 7.6 m with a pinwheel pattern around columns. Excellent crack control is reported but it should be noted that a full 7 days of water curing was specified.

A unusual floor in terms of jointing is the warehouse floor constructed in England about 1990 where four 30x40 m bays totaling 4800 m^2 were placed without intermediate joints on a slip membrane using laser-controlled screeds (Smith, 1991). The floor was 180 mm thick, and the permanent joint around each bay consisted of two 180 mm high back-to-back steel channels with steel dowel bars inserted through them.

Rehabilitation of existing industrial floors with thin fully bonded SFRC overlays has also become accepted, particularly in the automotive industry (Smith and Wilde, 1996). Typically the surface is milled and sandblasted to remove the top 6–25 mm as necessary to expose sound uncontaminated concrete prior to placement of SFRC with 30–45 kg/m^3 of fibers. Proper matching of overlay joints with existing joints in the old concrete is important. When this becomes impractical square panels with joints at 3–4 m are recommended (Smith and Wilde, 1996).

7.6.4.5. Water-Retaining Structures

Resistance to cracking to minimize leakage under exterior exposure conditions involving substantial changes in temperature and degree of saturation of the concrete as the water level changes is the primary concern in most of these applications. An example is the SFRC used in 1984 to replace the extensively cracked 1909 concrete in the upstream face of the Barr Lake Dam in Colorado (Fig. 7.39, upper). The SFRC replacement mixture contained 47 kg/m^3 of 60 x 0.8 mm hooked wire fiber in 38 mm aggregate concrete with 260 kg/m^3 of cement and 46 kg/m^3 of fly ash placed 102 mm thick over an area of 46,000 m^2 (Rettberg, 1986) (Mass, 1997). This is an example of SFRC with relatively long fibers because of the large aggregate and low cementitious material content used to minimize shrinkage. The replacement lining was slip-formed and jointed into 3 × 3 m panels. A visual survey of the 10-year old replacement lining shows that it continues to provide good water retention with minimal cracking which was the objective of the repair (Fig. 7.39, lower) (Mass, 1997).

In Australia, water-retaining applications include reservoirs, water storage tanks and swimming pools (Fig. 7.40) (Destree, 1995) (McGregor, 1995). In regions where water storage is a high priority, 20,000 litre precast tanks represent a significant application serviced for many years by SFRC typically with 75 kg/m^3 of enlarged-end slit sheet fibers in a thickness

FIGURE 7.39   Original lining to Barr Lake Dam before replacement (upper) and SFRC replacement lining after 10
years (lower) (Courtesy of Gary R. Mass)

tapered from 80 mm to 35 mm. Similar cast-in-place tanks can reach 6 m in diameter and accommodate up to 100,000 litres (Fig. 7.40) (McGregor, 1995). Swimming pools built with SFRC, 75–100 mm thick for cast-in-place construction and as little as 20 mm thick precast, are also common (Fig. 7.40). The issue of fiber content has become controversial in Australia as the European practice of using 20–40 kg/m^3 of relatively long 50–60 mm fibers comes into conflict and competition with the Australian practice of using 50–75 kg/m^3 of 18–25 mm long slit sheet fibers (McGregor, 1995).

7.6.4.6. Shotcrete

Since the 1970's steel fibers have been widely used in both dry and wet-process shotcrete for a wide variety of applications involving both new construction and repair (Henager, 1981–1) (ACI Committee 506, 1984). Steel fibers make possible the omission of the mesh reinforcement traditionally used in shotcrete, resulting in significant layer thickness reductions and cost savings along with improved performance, notably with respect to durability since the discontinuity of fibers tends to inhibit the classic galvanic corrosion process that occurs more easily when the reinforcement is electrically continuous (Hoff, 1987) (Morgan, 1994).

Slope stabilization, tunnel linings, or linings to other natural or manufactured containment devices including underground storage facilities and above-ground domes or similar thin-section structures, are among the applications in new construction where both dry and wet processes have been used, in some cases with robot-controlled placement techniques (ACI 506, 1984) (Morgan and McAskill, 1984) (Sandell and Westerdahl, 1986) (Opsahl, Buhre and Hornfeldt, 1986) (Vandewalle, 1990). SFRC has also been used extensively in shotcrete repairs to tunnels, bridges, culverts, buildings and marine structures damaged by freezing and thawing, chloride in seawater or deicing salts and other causes (Gilbride,

FIGURE 7.40   Australian use of SFRC in irrigation reservoirs
(upper), water storage tanks (middle) and precast or
cast-in-place swimming pools (lower) (Courtesy of
BHP Reinforcing Products)

Morgan and Bremner, 1988) (Johnston and Carter, 1989) (Morgan, 1994).

During the 1970's and early 1980's the dry process dominated projects in North America, while in Europe the wet process received increasing attention. Perceived advantages of the wet process include better control of water-cement ratio, less fiber loss due to rebound, a less dusty working environment and greater production capability for high-volume applications. The advent of superplasticizing admixtures helped facilitate the preparation of high-workability, pumpable mixtures needed for wet process application (Skatun and Spigerverk, 1986). However, the dry process is often preferable for low-volume work or intermittent placements typical of repair work, although the wet process can be adapted to intermittent operation using the so-called thin stream system (Morgan, 1994). High speed photographic examination of a dry-process shotcrete stream showed many fibers in the outer portion of the airstream being blown radially away from the intended target rather than rebounding from the surface, and the recommendations to reduce rebound include among other factors reduced air pressure and spraying at the wettest stable consistency (ACI Committee 506, 1984). Obviously, the wet process facilitates the latter. There is no consensus on whether the amount of fiber lost is proportionately greater or less than the total loss for the shotcrete as a whole in either process, but comparative data suggest that the lowest achievable rebound loss with either process is 10–20% and that it can often be much higher for the dry process (ACI Committee 506, 1984) (Banthia, Trottier, Beaupré and Wood, 1994). Clearly, the in-place toughness or residual strength of the SFRC is affected by fiber loss due to rebound.

One of the other factors recommended for reducing rebound (ACI Committee 506, 1984) is the inclusion of more fines, and in this respect the increasingly wide use of silica fume that started in the 1980's is very advantageous. Greater resistance to both adhesive and cohesive failures during application, less rebound, less need for accelerators, and a stronger and more impermeable matrix are its main advantages. Silica fume with steel fibers tends to produce a shotcrete that combines the advantages of the fibers in terms of toughness and residual strength with the advantages of ease of application and a strong impermeable matrix associated with silica fume (Morgan, 1988) (Johnston and Carter, 1989). Steel fiber reinforced silica fume shotcretes typically comprise 350–425 kg/m^3 of cement with 5–10% silica fume by weight of cement and 40–80 kg/m^3 of steel fibers.

In earlier projects up to the late 1980's the SFRC for shotcrete applications tended to be specified in terms of flexural strength, supposedly to reflect the benefit of including fibers, as well as compressive strength, absorption and permeable voids to confirm the quality of the matrix. Subsequently, recognition of the importance of the ability of the SFRC to withstand the large deflections produced by earth and rock movements and impacts due to falling rock in tunnelling led to specifications incorporating toughness or residual strength parameters instead of or in addition to flexural strength. In North America this typically took the form of minima for ASTM C 1018 toughness indices I_5, I_{10}, and I_{30} (Table 7.2) (Morgan, 1991) (Lewis, 1993) (Forrest et al., 1995).

A total of 21 applications of dry or wet-process steel fiber shotcrete is listed for Canada between 1980 and 1990 (Morgan, 1991). The seismic retrofit of the Littlerock dam in California in 1994 is an example of a large-scale application of wet-process shotcrete applied about 125 mm thick to 4800 m^2 of a multi-arch dam face using 60 kg/m^3 of hooked wire fibers in a superplasticized air-entrained matrix with 405 kg/m^3 of cement and 41 kg/m^3 of silica fume (Fig. 7.41, upper) (Forrest et al., 1995). The 1.25 km twin two-lane highway tunnels connecting the Kentucky and Tennessee portions of US 25E at Cumberland Gap are another example of a large-scale application of wet-process shotcrete for both the tunnel lining and the rock slope stabilization in the approaches using 53 kg/m^3 of

TABLE 7.2 Typical shotcrete specifications for SFRC

Project identification	Age days	Strengths — MPa		Toughness indices		
		Compression	Flexure	I_5	I_{10}	I_{30}
Ref. Morgan, 1991	28	40	6.0	3.5	5.0	14.0
Cumberland Gap	7	32*	3.5	4.0	6.0	N.R.
Littlerock Dam	28	41	4.8	3.5	5.0	16.0
Alberta Bridges	7	40	N.R.	N.R.	N.R.	N.R.

N.R. — No requirement   * At 28 days

hooked wire fibers in a superplasticized air-entrained matrix with 439 kg/m^3 of cement and 59 kg/m^3 of silica fume (Fig. 7.41, lower) (Lewis, 1993). As this project evolved the parties agreed to substitute for the original I_5 and I_{10} criteria (Table 7.2) a residual strength requirement at the I_{10} end-point deflection of 60% of the specified flexural strength in order to resolve difficulties created by improper testing that resulted in non-compliance for I_5 contrasting with compliance for $I_{10.}$

The repairs to numerous Alberta highway bridges started in 1984 are an example of smaller-scale dry-process shotcrete using 60 kg/m^3 of hooked or crimped 25–30 mm long steel fibers often with silica fume (Johnston and Carter, 1989). Prebagged material is often used on these smaller-scale projects that usually involve repair of piers, abutments and precast girders damaged by freezing and thawing with deicing salt. Repairs to the North Saskatchewan River bridge done in 1986 are typical, and were reported essentially crack-free in 1990 (Morgan, 1994).

Other examples of successful use of SFRC in infrastructure restoration include wet-process repairs to marine wharves in St. John, New Brunswick, and dry-process repairs to transmission tower caissons in British Columbia and the upstream face of the Jordan River dam on Vancouver Island (Morgan, 1994). The exposure condition at the St. John wharves is particularly severe in the intertidal zone because of the 8.5 m tide range combined with 200–300 cycles of freezing and thawing per year. The wet-process SFRC placed 100–150 mm thick in 1986 using 60 kg/m^3 of fibers with silica fume has performed well through an estimated 1000 cycles of freezing and thawing, and continues to be used in subsequent wharf repairs (Gilbride, Morgan and Bremner, 1988) (Morgan, 1994).

In the 1990's a trend towards shotcrete specification criteria based on higher end-point deflections has developed mainly as a result of recognizing the need to eliminate the possible effects on test results of unstable strain-softening and inconsis-

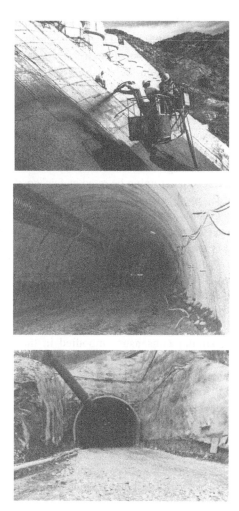

FIGURE 7.41   Wet process steel fiber shotcrete for seismic retrofit
of Littlerock multi-arch dam (upper) (Courtesy
AGRA Earth & Environmental, Ltd.) and highway
tunnel construction at Cumberland Gap (middle and
lower) (Courtesy of Bekaert Corporation)

tencies in determining first crack, as discussed in 7.2. In Europe the criteria in 1990 varied by country between ASTM C 1018 $R_{10,30}$ or $R_{10,50}$ residual strength factors and a residual strength factor R_e based on the JCI SF-4 equivalent flexural strength as a percentage of the ultimate flexural strength (Vandewalle, 1990). Numerous tunnelling projects are identified but project specifications are not given, although other data in the reference and elsewhere (Morgan, Mindess and Chen, 1995) suggest values of 70–100 for $R_{10,30}$ and 60–90 for R_e as being achievable. In 1988 it was suggested that shotcrete for tunnel lining be rated as fair, good or excellent according to minimum $R_{10,30}$ values of 40, 60 and 80 respectively (Vandewalle, 1990). Subsequent experience in the United States particularly at Cumberland Gap, as discussed in the previous paragraph, suggests that values in the 40 to 60 range may be more realistic for tunnel linings. Residual strength requirements based on equivalent flexural stress retained at deflections in the range 0.5 to 4 mm are used instead of residual strength factors in the 1993 Norwegian guide (Norwegian Concrete Association, 1993) and the broader but similar consensus embodied in the European Specification for Sprayed Concrete (EFNARC, 1996) illustrated previously (Fig. 7.11). Residual strengths in these specifications for a 5 MPa flexural strength using a $450 \times 125 \times 75$ mm beam vary from 10% to 90% of the flexural strength according to performance class and end-point deflection. The EFNARC specification also incorporates an option based on the energy needed to load a two-way slab to 25 mm deflection as required by the French railway company (See Section 7.5).

In Canada, the trend is similar with one approach being to specify residual strength limits expressed as percentages of a design flexural strength instead of the actual flexural strength of test specimens. It was proposed for classifying shotcrete performance according to load-deflection relationships obtained using $300 \times 100 \times 100$ mm beams (Morgan, Mindess and Chen, 1995) (Fig. 7.13). Performance is classified in four levels according to the design flexural strength, which is primarily matrix dependent, and the post-crack residual strength which is primarily dependent on the fiber geometry and amount present after rebound. This system facilitates identification of wide differences in performance attributable to changes in the shotcreting process, nominal fiber content (before rebound) and fiber geometry (Table 7.3) (Morgan, Chen and Beaupré, 1995).

It was used to evaluate the wet-process shotcrete in test panels simulating the vertical and inclined portions of the Littlerock dam retrofit (Table 7.2). Based on a design flexural strength of 4.8 MPa, performance levels III and IV (Fig. 7.13) were achieved for the vertical and inclined panels respectively with 60 kg/m³ of hooked 30x0.5mm fibers in a superplasticized, air-entrained matrix with 405 kg/m³ of cement, 41 kg/m³ of silica fume and an in-place air content of 5 ± 1% (Forrest et al., 1995).

7.6.4.7. Structural and Precast

SFRC has been used or investigated for a wide variety of specialized applications or potential applications too numerous to describe in detail. They include dolosses for ocean wave protection, vaults and safes, mine crib blocks, tilt-up panels and precast garages (ACI 544, 1996), hollow concrete electrical poles (Rotondo and Weiner, 1986), and storage tanks for nuclear waste (Pech and Schacher, 1992). Precast segments 150 to 300 mm thick assembled to form 3–6 m diameter tunnel linings have been used in several European proejcts in Italy and the U.K. for water, service and underground rail tunnels. A similar project for the Cigar Lake mine development in Saskatchewan started early in 1998 (Bekaert Corp., 1996). Proposed structural uses in connections subjected to seismic loading (Katzensteiner et al., 1994), replacement of stirrup reinforcement for shear in beams (Batson and Youssef, 1994), and repair of damaged portions of reinforced beams

TABLE 7.3 Performance levels for a range of steel fiber shotcretes

Shotcrete process	Fiber details			Flexural strength MPa	Performance level for design flexural strength	
	Type	Size mm	Amount kg/m³		4.0 MPa	5.0 MPa
Wet	Slit sheet	25	53	7.2	I	I
Wet	Mill-cut	32	54	7.1	II	II
Wet	Crimped	32	52	6.5	II	II
Wet	Hooked wire	28 × 0.5	49	6.3	IV	III
Wet	Hooked wire	30 × 0.5	60	6.4	IV	IV
Dry	Hooked wire	30 × 0.5	60	6.3	III	III
Dry	Hooked wire	30 × 0.5	60	5.5	III	II
Dry	Mill-cut	32	60	7.3	III	II
Wet	Hooked wire	30 × 0.5	60	4.6	IV	N/A
Wet	Crimped	38	60	4.2	III	N/A

(Andrews and Sharma, 1990) have received considerable study, but have not yet been widely exploited in applications (Balaguru, 1995) (Craig, 1987) (ACI 544, 1988–2).

7.6.4.8. *Slurry Infiltrated Steel Fiber Concrete (SIFCON)*

The slurry infiltration process described previously in 5.2.4 permits fiber contents much greater than the upper limit of about 2% by volume achievable in conventionally mixed SFRC, typically 4–12% in practice (Fig. 7.42, left) (Naaman, 1992), although as much as 27% has been reported in laboratory trials (Naaman, Otter and Najm, 1991). Compared to conventionally mixed SFRC with 2% fibers, the degree of property enhancement possible in SIFCON is much greater with, for example, flexural strengths of 60 MPa using 14% of 30x0.5 mm hooked wire fibers. The initial linear elastic portion of the load-deflection relationship up to matrix cracking is greatly extended and the levels of toughness and

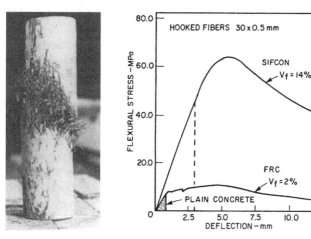

FIGURE 7.42   Cylinder compression failure of SIFCON (upper) (Courtesy of A.E. Naaman) and comparative flexural performance of SIFCON, conventional SFRC and the parent matrix (lower) (Naaman, 1992)

residual strength retained to large deflections are also extremely high (Fig. 7.42, right) (Naaman, 1992). Strength and toughness in direct tension and shear are also substantially improved, while in compression toughness rather than strength is primarily affected (Naaman, 1992) (van Mier and Timmers, 1992) (Naaman, Otter and Najm, 1991).

The SIFCON manufacturing process tends to favor thin-section applications because ensuring complete penetration of the fibers by the slurry matrix becomes more difficult as thickness increases. Several thin-section applications and their performance in service are reported (Krstulovic-Opara et al., 1995). Precast SIFCON panels 50 mm thick placed over a newly constructed concrete subbase at an airport parking gate to support the main wheel carriage of a Boeing 727 showed no cracking after 1 year of service (Lankard, 1986). Precast SIFCON panels 25 mm thick used to protect the floor from falling pieces in a metal processing yard solved the problem of frequent replacement of floor slabs (Mondragon, 1984). A 15x5.5 m deteriorated area of concrete in a parking lot repaired with a 25 mm thick cast-in-place SIFCON bonded overlay without joints was reported in a good condition after 9 years exposure to traffic and deicing salts (Mondragon, 1984) (Schneider, 1992). A 25–50 mm SIFCON repair to a bridge deck topped with 10 mm crushed rock to improve wear resistance was reported free of cracking, abrasion and fiber corrosion after 7 years (Schneider, 1992). A precast prestressed bridge girder damaged by an overheight truck impact was repaired and returned to service in four days in circumstances where replacement of the girder and a part of the deck were estimated at a much higher cost with a 3-month closure of the bridge required (Schneider, 1992). Various military applications involving explosive storage cabinets, barriers to resist small arms ballistics, and grenades or mortar fire, all of which exploit the great toughness and impact resistance of SIFCON are also reported (Schneider, 1992). Uses in transportation containers for radioactive waste, security vaults, blast-resistant doors and joint systems for precast building elements are under consideration (Schneider, 1992). Uses in beam-column connections, the lower sections of shear walls or lower columns in high rise buildings, the punching shear zones in slab-column systems, and the highly stressed zones around prestressing anchors are proposed for improving resistance to earthquake effects (Naaman, 1992).

The matrix in most of the applications consists of cement with 25–35% fly ash, superplasticizer, and a water-cementitious material ratio of 0.25–0.32 (Schneider, 1992). Fiber content is 10–15% by volume. After evaluating various fiber-matrix systems in terms of performance and cost one fiber manufacturer recommends 8.5–10.5% by volume of 60x0.8 mm hooked wire fibers in mortar with maximum particle size 0.5 mm and a 28-day strength of 60 MPa, giving SIFCON 28-day strength minima of 25 MPa and 85 MPa in flexure and compression respectively (Vandenberghe, 1992).

7.7. POLYMERIC FIBER-REINFORCED CONCRETE (PFRC)

Most of the fibers other than metal, glass and natural fibers that are suitable for use in concrete in terms of being compatible with the alkaline environment and conventional mechanical mixing processes are in fact polymers. They include polypropylene, polyethylene, nylon and polyester. Such fibers for fiber-reinforced concrete or shotcrete are also classified as synthetic, and according to ASTM C 1116 (ASTM, 1994) are subject to proof of compatibility with moisture, cement alkalis and chemical admixtures, except for homopolymer polypropylene which is identified as satisfactory in these respects. The compatibility of other polymeric fibers with the moist alkaline environment in cement paste has already been discussed in the context of fiber-reinforced cements in 6.8, and the same considerations apply to fiber-reinforced concretes.

Polypropylene has tended to be the most widely used polymeric form of fiber reinforcement in concrete because of its excellent resistance to moisture, acids and alkalis and the cheapness of the raw material on a volume basis compared with steel and other alternatives (Krenchel and Jensen, 1980) (Larsen and Krenchel, 1991). The fibrillated form of polypropylene, developed to augment the intrinsically poor adhesive bond of the monofilament form by mechanical bonding due to separation and branching of the fibrils in the strand as discussed previously in 3.2.2, is most widely used in concrete. Monofilament polypropylene is also available, as are monofilament forms of polyethylene, polyolefin, nylon and polyester, sometimes with proprietary surface treatment or surface texturing to improve bond and therefore pullout resistance and overall reinforcing effectiveness (Krenchel and Shah, 1985) (Portland Cement Association, 1991).

7.7.1. Mechanical Properties

The amount of polymeric fibers used in concrete varies widely depending on the intended purpose of including fibers and the type and geometry of the fibers. A minimum of 0.1% by volume (0.9 kg/m^3 for polypropylene of density 910 kg/m^3) is widely used in flatwork with the intention of reducing or eliminating the possibility of plastic shrinkage cracking within the first few hours after placement. Whether this amount is sufficient to replace welded wire mesh for control of later-age cracking in the hardened concrete is controversial (Zollo and Hays, 1991). Greater amounts ranging from 0.5–0.7% for fibrillated polypropylene to 1.5% for monofilament polyolefin are generally used when significant later-age improvement in mechanical properties after hardening is intended.

7.7.1.1. Slow Flexure

Performance is most commonly evaluated in bending, as in the case of SFRC, and often for direct comparison with SFRC.

Naturally, fiber content is a dominant factor affecting performance. However, fiber contents in the range 0.1 to 0.7% have no significant effect on flexural strength for low-modulus polypropylene fibers (Berke and Dallaire, 1994) (Ramakrishnan et al., 1994) (Zellers and Ramakrishnan, 1994) (Malhotra, Carette and Bilodeau, 1994) (Banthia and Dubey, 1996), and even for high-modulus polyethylene (Soroushian, Khan and Hsu, 1992). Their primary effect is on the post-crack portion of the load-deflection relationship in terms of toughness and residual strength. Improvements in these properties are small at fiber contents of the order of 0.1% by volume, but become increasingly significant as the fiber content increases. The load-deflection behaviour is typically characterized by rapid strain-softening after first crack followed by plastic behaviour with the residual load remaining essentially constant to large deflections (Fig. 7.43, upper) (Banthia and Dubey, 1996).

Since the residual load remains essentially constant over a wide range of deflection, it is reasonable to compare the performance obtained with different amounts and types of fibers in terms of the various ASTM C 1018 residual strength factors reported by different investigators, provided of course they represent the residual load over a deflection range large enough not to be affected by unstable strain-softening as previously discussed in 7.2.2. A compilation of such data involving residual strength factors $R_{10,20}$, $R_{10,30}$, $R_{10,50}$ and $R_{30,50}$ illustrates the effects of fiber amount, length, configuration (monofilament or fibrillated), and to a lesser extent type (polypropylene vs. nylon) on post-crack flexural performance (Fig. 7.43, lower) (Banthia and Dubey, 1996) (Johnston, 1995–1) (Chen et al., 1995) (Malhotra et al., 1994) (Morgan et al., 1992) (Zhang et al., 1997). Some of the data are averages of up to six tests involving either the same mixture tested in different laboratories or the same fiber tested in different matrices in a single laboratory.

Despite differences in testing arrangements, matrix characteristics and matrix strength, the effect of increasing fiber

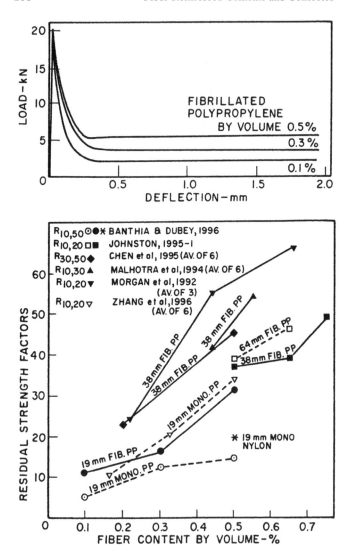

FIGURE 7.43   Flexural load-deflection relationships for FRC with
19 mm fibrillated polypropylene (Banthia and
Dubey, 1996) (upper), and residual strength factors
for monofilament and fibrillated polypropylene and
nylon (lower) (sources identified in figure)

content is clear enough to provide an idea of the order of magnitude of residual strength factor to be expected at various fiber contents (Fig. 7.43, lower). The effect of increasing fiber length is also fairly clear, at least in terms of the benefit of changing it from 19 to 38 mm, but any benefit of additional length beyond 38 mm is slight as mixture stiffening and consolidation difficulty tend to counteract any increase in reinforcing effectiveness (Johnston, 1995–1). The fibrillated form of polypropylene appears slightly superior to the monofilament form of the same length (Banthia and Dubey, 1996). However, differences in performance between the monofilament and fibrillated forms and between polypropylene and nylon are small in this data compilation (Fig. 7.43, lower). Logically, aside from the primary effect of fiber content, performance should depend to some extent on the final aspect ratio of the separated fibrils or monofilaments after mixing. At low fiber contents performance differences between polypropylene and other polymeric alternatives like nylon or polyester are probably small, but may increase at higher fiber contents. Unfortunately, there is a serious lack of data for fibers other than polypropylene in the refereed technical literature.

Differences in matrix strength may also contribute to differences in residual strength factor, particularly when matrices with a high percentage of fly ash are involved, as in the data of Malhotra et al., 1994, Morgan et al., 1992 and Zhang et al., 1997 (Fig. 7.43, lower). Low matrix strength associated with early-age testing or high fly ash content causes temporarily higher residual strength factors until the matrix matures (Malhotra et al., 1994) (Zhang et al., 1997), so the data shown in Fig. 7.43 are limited to mature matrices with flexural strengths greater than 4 MPa and generally not more than 6 MPa. This trend of decrease is residual strength factor with increasing matrix strength is consistent with the corresponding trend for steel fibers, as discussed previously in 7.6.1.1.

Compared with the relatively small size and high aspect ratio of fibers typical for monofilament polypropylene, for example 19x0.056 mm of aspect ratio 340 (Zhang et al., 1997), and the similarly small fibrils of uncertain size that result from the separation of fibrillated polypropylene, the fibers identified as polyolefin that are packaged in paper-wrapped bundles as described previously in 5.2.2.3 (Fig. 5.16), are larger and of lower aspect ratio, typically 50x0.63 mm (aspect ratio 80) for precast or cast-in-place construction and 25x0.38 mm (aspect ratio 67) for shotcreting (Morgan, 1995) (Morgan and Rich, 1996). The lower aspect ratio and specific surface of these fibers raise the threshold fiber content at which mixture stiffening becomes excessive beyond the 0.5–0.6% for the finer fibers to as much as 1.25–1.5% depending on fiber aspect ratio and method of application. Residual strength factors are substantial and can exceed 50 for workable mixtures (Fig. 7.44) (Morgan, 1995).

7.7.1.2. *Flexural Impact*

Qualitatively, there is consensus that polymeric fibers improve impact resistance, as in the case of steel fibers, but most of the supporting data are based on the simplistic empirical drop-weight test (ACI 544, 1988–1) where the net energy input to the specimen is not determined. As little as 0.1% polypropylene appears to significantly improve performance in this rather questionable test, the number of blows to first crack for mixtures with 0.1% of 19 mm monofilament polypropylene fibers being 2–3 times the number for the control mixture without fibers (Ramakrishnan et al., 1994).

Quantitatively, impact energy determined using the weighted-pendulum technique (Hibbert and Hannant, 1981) is approximately equal to the energy measured for slow flexural loading using 100 mm square beams tested to 10 mm deflection, consistent with the effect of steel fibers on impact energy (Fig. 7.26). Impact energy improvements relative to the matrix without fibers vary with fiber content, fiber length, matrix composition or strength,and probably testing configuration. Data compiled from weighted-pendulum tests on con-

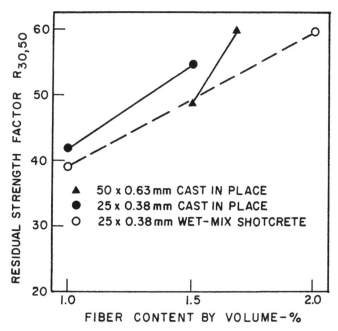

FIGURE 7.44   Residual   strength   factors   for   monofilament
                polyolefin (Morgan, 1995)

cretes with 35 or 75 mm fibers and a relatively strong (w/c =
0.40) matrix (Hibbert and Hannant, 1981) and from drop-
weight tests on concretes with 19 mm fibers and a relatively
weak (w/c = 0.79) matrix (Mindess, Bentur, Yan and Vondran,
1989) (Mindess and Vondran, 1988) show a fairly consistent
and significant improvement in impact resistance with increase
in fiber content (Fig. 7.45). However, the almost 60%
improvement attributed 0.1% by volume of 19 mm fibers is
questionable, and could be due to imprecise determination of
the relatively low fracture energy for the unreinforced matrix.
On the other hand, the rate of improvement in impact energy
from 0.1% to 0.5% of 19 mm fibers, as depicted by the slope

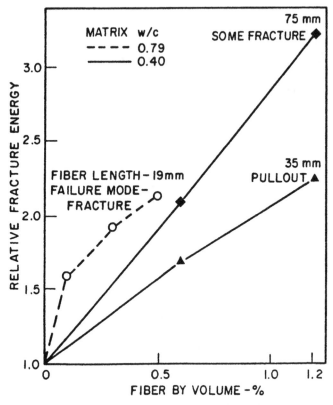

FIGURE 7.45   Effect of fiber content, fiber length and matrix composition on flexural impact resistance of concrete
with fibrillated polypropylene fibers (Hibbert and
Hannant, 1981) (Mindess et al., 1989)

between 0.1 and 0.5% fiber content, is consistent with the rate
of improvement for the 35 mm fibers.

Tests using the 19 mm fibers in combination with 1.60% of
conventional steel reinforcement in the tension zone of the
beams tend to confirm a more modest improvement in impact
energy at the 0.1% fiber content, 22% compared to the almost

60% without rebar (Fig. 7.46, upper). Load-deflection relationships for the reinforced beams with 19 mm fibers in drop-
weight impact tests illustrate the increased energy absorption
capability associated with up to 0.5% of fibers (Fig. 7.46,

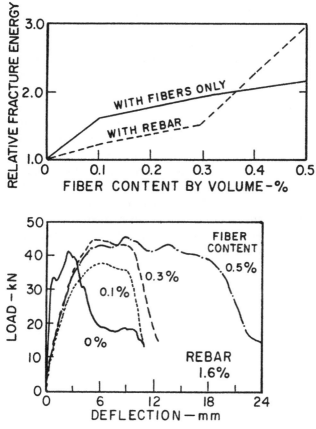

FIGURE 7.46   Drop-weight impact performance of beams with
1.60% of conventional steel reinforcement and
various amounts of 19 mm fibrillated polypropylene
fibers in terms of relative fracture energy (upper) and
load-deflection behavior (lower) (Mindess et al.,
1989)

lower) (Mindess et al., 1989). This suggests that fibers might enhance the performance of conventionally reinforced members subjected to earthquake loading.

Any comparison of performance between impact and slow flexure involving composites reinforced with polypropylene fibers is also probably influenced by the relatively high sensitivity of polypropylene to strain rate. For example, polypropylene fibers are reported to exhibit an elastic modulus under impact loading of 2–3 times the normal value for static loading (Zonsveld, 1975). This tends to enhance the performance of polypropylene over other less rate-sensitive fibers like steel in any test that produces a high strain rate. Since the impact energies for the conventionally reinforced beams are typically at least 10 times the values for the beams with fiber only, the same impacting weight and drop height seem likely to produce much lower strain rates in the reinforced beams than in the relatively low-energy beams with fiber only. As in all impact work, the possibility of variation in strain rate between different testing arrangements, and between differently reinforced specimens tested with the same impacting device and energy input, makes interpretation of impact data difficult.

7.7.1.3. Flexural Fatigue

The different ways of interpreting fatigue data previously discussed in 7.6.1.3. for SFRC also apply to polymeric fiber-reinforced concrete. In addition, since the available data relate to polypropylene which is strain-rate sensitive, it is likely that the faster the rate of cycling the better the performance because the time available for time-dependent viscoelastic stretching of the fibers is minimized. In tests that evaluated steel and polypropylene fibers (Ramakrishnan, Wu and Hosalli, 1989), the percentages of static ultimate flexural strength sustainable to 2 million cycles for 0.5 and 1.0% of 19 mm fibrillated polypropylene are mainly in the 65–70% range, close to the 65% reported for the control concrete without fibers. Results from the same laboratory also for 19 mm fibrillated polypropylene indicate a slight increase in the 2 million cycle endurance limit with increase in fiber content. These values of 58, 59 and 69% for fiber contents of 0.1, 0.5 and 1.0% respectively (Nagabhushanam, Ramakrishnan and Vondran, 1989) are within the 55–70% range obtained for low strength steel fibers and concrete without fibers (Fig. 7.29). Subsequent tests with 19 mm monofilament polypropylene at fiber contents of only 0.067% and 0.1% also show no significant difference from the control concrete without fibers (Ramakrishnan et al., 1994).

Unfortunately, no data have been located that indicate the fatigue performance of polymeric fibers other than polypropylene. It is possible that high-strength high-modulus polymeric fibers may enhance fatigue performance in much the same way as the higher strength steel fibers (Fig. 7.29).

7.7.2. Resistance to Cracking

The widespread use of relatively small amounts of polypropylene, polyester and nylon fibers, typically 0.1% by volume, primarily to control plastic shrinkage cracking, has led to interest in the effect of fibers on early-age cracking of the freshly placed concrete as well as on later-age cracking of the hardened concrete associated with drying shrinkage or temperature decrease under conditions of restraint.

7.7.2.1. Early-Age Cracking

Many factors unrelated to fibers contribute to plastic shrinkage and the possibility of cracking associated with it (Shaeles and Hover, 1988). Despite the possibility that proper mixture proportioning, screeding and finishing techniques and curing procedures can prevent plastic shrinkage cracking in most situations, a fiber content of about 0.1% is often advocated as an effective remedy for difficult concreting conditions involving high evaporation rates. The procedure for evaluation of plastic shrinkage cracking potential usually involves blowing air over the surface of a freshly placed concrete slab that is restrained at or near its edges, and quantifying cracking as the product of crack length and width per unit area of tested surface for comparison with cracking of a corresponding control concrete without fibers (Kraai, 1985) (Berke and Dallaire, 1994) (Soroushian, Mirza and Alhozaimy, 1995) (Balaguru, 1994). A version of this procedure is being considered for adoption as an ASTM standard. Typically, the results demonstrate the effect of fibrillated polypropylene fiber content and length on reducing the severity of plastic shrinkage cracking relative to a control concrete without fibers (Fig. 7.47, upper) (Berke and Dallaire, 1994). Less commonly, they indicate the actual crack area and maximum crack width (Fig. 7.47, lower) (Soroushian, Mirza and Alhozaimy, 1995). In the latter case, the influence of different screeding rates and finishing procedures was investigated, but their effects appear minimal for concretes with fibers and only significant for plain concrete, with faster screeding rates and lack of finishing apparently increasing both crack area and crack width.

Other types of fiber have been similarly compared using steel, nylon, polyethylene, polypropylene, polyester and cellulose fibers of normal length and microfibers made from pulp (Balaguru, 1994). The results for 19 mm nylon monofilaments demonstrate the expected influence of fiber content with 0.1% by volume reducing cracking by about 75% (Fig. 7.48, upper) (Balaguru, 1994). This is comparable to the 70% reduction for 0.1% of 19 mm fibrillated polypropylene (Fig. 7.47, upper), although the matrices and test conditions are not identical. The comparison of 19 mm fibrillated polypropylene and 19 mm nylon and polyester monofilaments at 0.067% by volume in a rich mortar matrix with high shrinkage potential shows nylon superior to polypropylene and polyester, attributed mainly to higher fiber count per unit of volume (Fig. 7.48, lower). Differences in aspect ratio, found to be important in both the tests with steel fibers and the tests with microfiber pulp, are

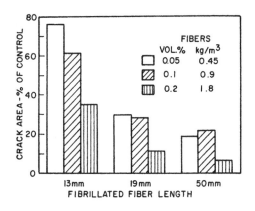

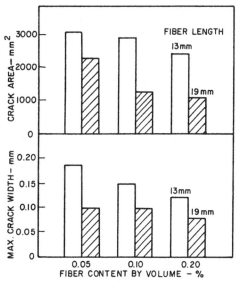

FIGURE 7.47   Effect of polypropylene fiber content and length on plastic shrinkage reduction under passage of air at 7.6 m/s, 35°C and 40% relative humidity (Berke and Dallaire, 1994) (upper), and under passage of air at 3.6 m/s, 24–27°C and 50% relative humidity (Soroushian, Mirza and Alhozaimy, 1995) (lower)

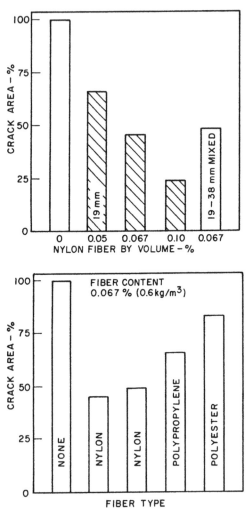

FIGURE 7.48   Effect of fiber content (upper) and fiber type (lower)
on plastic shrinkage reduction under passage of air at
5.4 m/s, 22°C and 50% relative humidity (Balaguru,
1994)

not discussed for the polymeric fibers in the published report (Balaguru, 1994).

Alternatives to the restrained slab test that have been used to demonstrate the early-age crack-reducing effect of small amounts of fiber include a slab with a restraining steel ring inserted at the centre (Padron and Zollo, 1990), and a form of ring test using a 280 mm diameter wood core with an outer steel ring having radially attached steel ribs penetrating 30 mm into the 150 mm radially wide specimen to provide restraint (Dahl, 1985). Both these alternatives have demonstrated that 0.1% fibers can significantly reduce plastic shrinkage cracking, typically by at least 75%, i.e. to 25% of the control. In slab tests, two forms of 19 mm polypropylene and a 10 mm acrylic fiber were compared, and the acrylic fiber was reported as more effective (Padron and Zollo, 1990). In ring tests, fibrillated polypropylene was tested in lengths of 13, 19 and 38 mm, but no significant differences in performance attributable to length were noted (Dahl, 1985). In subsequent similar ring tests, cracking is reported to depend strongly on cement type as well as on fiber type (monofilament versus fibrillated), but the scatter of the results was large enough to prevent firm conclusions on the effect of some fiber variables (Vandewalle, 1997).

Regardless of the many differences in matrices and testing techniques, it is clear that small amounts of the order of 0.1% by volume of fibers can significantly reduce cracking in matrices with a high plastic shrinkage potential. Cracking of this type occurs within the first 6 to 8 hours after casting, and does not increase thereafter under normal storage conditions which are of course much less severe than the forced-air conditions in the various test procedures.

7.7.2.2. Later-Age Cracking

The ring form of restrained shrinkage test originally developed for evaluating the effect of steel fibers in concrete, as described previously in 7.6.2, can also been used to evaluate the effect of polymeric fibers on later-age resistance to cracking caused by drying. Results for polypropylene fibers using specimens dried at 20°C and 40% relative humidity after 4 days of moist curing indicate that the 0.1% of fibers that effectively reduce plastic shrinkage cracking (Fig. 7.47) have little effect on drying shrinkage cracking (Fig. 7.49, left) (Grzybowski and Shah, 1990). However, a larger fiber amount of 0.5% or more significantly reduces the maximum crack width (single crack only at 0.1, 0.25 and 0.5% fibers) and may promote multiple cracking (two cracks at 1.0% fiber content). Polypropylene fibers in fibrillated form 19 mm long were comparatively less effective than 25 mm long steel fibers in reducing the average crack width (Fig. 7.49, right). Using the same testing technique with drying started at 4 hours instead of 4 days, the maximum crack width for 0.5% of 19 mm polypropylene was about the same as for 0.25% of 30 mm hooked steel wire (Shah, Karaguler and Sarigaphuti, 1992). In other work using ring specimens with five kinds of polypropylene fibers at 0.75% by volume, differences attributable to fiber length were

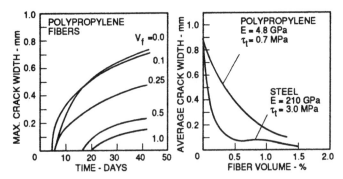

FIGURE 7.49   Effect of fiber content on shrinkage cracking during drying for 19 mm fibrillated polypropylene (left), and after 6 weeks of drying for 19 mm polypropylene and 25 mm steel fibers (right) (Grzybowski and Shah, 1990)

discernible with 38 mm long fibers producing less severe cracking than 12 or 19 mm lengths (Vandewalle, 1997).

7.7.3. Durability of Polymeric Fibers in Concrete

Many of the issues of compatibility of fibers with the moist alkaline environment in cement paste have been mentioned in the context of fiber-reinforced cements in Chapter 6. Only the fiber types seriously considered or frequently used in concrete are discussed further, namely polypropylene, polyethylene, polyester, nylon and acrylic.

While durability in concrete is in some respects specific to the chemistry of each fiber type, some general physical considerations can be important. All these polymers melt at a relatively low temperature between about 134°C for polyethylene and 257°C for polyester (ACI 544, 1996), so they cannot be expected to perform under conditions where the concrete temperature approaches or exceeds these values, as in the case of autoclave curing during manufacture or fire in service. For example, it has been shown that polypropylene fiber-reinforced concrete autoclaved at 140°C and then oven dried at 116°C for 24 hours suffered considerable loss of ductility due to oxidative degradation of the fibers (Mai, Andonian and Cotterell, 1980). Separately, it has been shown that both fiber and composite concrete strengths using polypropylene fibers decrease sharply at temperatures above about 120°C, about 75% of the 165°C melting point of polypropylene (Krenchel, 1987). Accordingly, the possibility of fiber and composite degradation should be considered for any polymeric fiber likely to be exposed to temperatures above about 75% of its melting point.

All polymeric fibers have a high specific surface. Even in small amounts of the order of 0.1% by volume they cause significant mixture stiffening which may reduce settlement of the aggregate particles thus reducing bleeding and the formation of bleed channels (Zollo, Ilter and Bouchacourt, 1986)

(Soroushian, Mirza and Alhozaimy, 1995) (ACI 544, 1996). However, there is no evidence of any resulting decrease in permeability for uncracked concrete. Indeed, chloride permeability tests using the rapid electrical method showed that up to 0.55% (5 kg/m^3) of polypropylene has no significant effect on the amount of charge passed, which depends primarily on matrix parameters such as water-cement ratio and presence of fly ash or silica fume (Malhotra, Carette and Bilodeau, 1994). Pressurized water permeability tests on specimens moist-cured prior to testing showed no significant change in permeability as a result of inclusion of 0.2% by volume of polypropylene fibers (Al-Tayyib and Al-Zahrani, 1990). In contrast, pressurized air permeability tests on specimens dried at 40°C for 24 hours before testing showed significantly reduced permeability for 0.3% polypropylene fibers with minimal change for 0.1% fibers in mortars clearly microcracked by drying (Sanjuan et al., 1991). Accordingly, fibers may reduce the permeability of cracked concrete to the extent that they reduce cracking due to plastic or drying shrinkage.

The chemical compatibility of fibers with the alkalis in cement, discussed in Chapter 6, indicated the uncertain or questionable durability of aramid and acrylic fibers in fiber-reinforced cements (Wang, Backer and Li, 1987). In contrast, nylon, polypropylene, and polyethylene fibers were reported as very resistant to strong alkalis, while polyester, like acrylic and aramid fibers, was reported as not resistant to some strong alkalis (Lyle, 1976) (Wang, Backer and Li, 1987).

More recent comparative tests of nylon, polypropylene and polyester in concrete with about 0.5% by volume (4.75 kg/m^3) of 19 mm fibers, involving accelerated aging in 50°C lime-saturated water following the practice widely used for fiber-reinforced cements, showed that concretes with nylon and polypropylene retain strength and toughness after 52 weeks accelerated aging, while concrete with polyester fibers exhibited significant losses of post-crack strength and toughness as illustrated by the load-deflection relationships (Fig. 7.50)

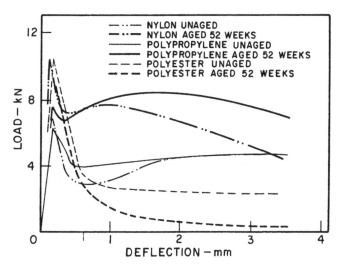

FIGURE 7.50   Effect of accelerated aging in 50°C water on load-
deflection relationships for nylon, polypropylene and
polyester fibers (Balaguru and Slattum, 1995)

(Balaguru and Slattum, 1995). The results are consistent with
other aging tests on polypropylene (Larsen and Krenchel,
1991) and on polyester (Jelidi et al., 1991).

The question of possible deterioration of polymeric fibers
in concrete as a consequence of the susceptibility of some
polymers to damage by ultraviolet radiation has also been
raised. While fibers within uncracked concrete are protected,
fibers bridging cracks in concrete could conceivably deteriorate due to the effects of ultraviolet light combined with moisture, as is commonly observed for rope. This possibility was
examined using a standard ASTM test for evaluating the deterioration of geotextiles, which subjects the fibers to ultraviolet
light at 65°C for up to 500 hours. Fiber strengths compared
before and after exposure show that nylon retained 95% of its
strength after 500 hours while one type of polypropylene
retained 63% of its strength and another type of polypropy-

lene disintegrated (Balaguru and Slattum, 1995). The possibility that some of the fibers may have contained additives to protect them from ultraviolet light is not discussed, but may have contributed to the difference between the two types of polypropylene fibers.

Obviously, deterioration with time is inconsequential for small fiber contents if they are intended only to mitigate the severity of plastic shrinkage cracking during the first few hours after casting. Only at higher fiber contents, where the intent is to improve the long-term properties of the hardened concrete, are alkali-related aging or ultraviolet-related deterioration likely to be of concern. In such cases, the possibility that fiber composition may vary within a generic group, such as polyesters, complicates any attempt to rate durability for the product group. Some fiber manufacturers claim that not all polyesters are subject to alkali attack. The attack mechanism is apparently hydrolysis and dissolution of polyethylene terephthalate in the presence of cement alkalis (Jelidi et al., 1991), but the distinction between polyesters that are subject to significant attack and those that are not, if any, is unclear.

7.7.4. Polymeric Fiber-Reinforced Concrete (PFRC) Applications

Applications involving polymeric fibers fit technically into two categories. The first involves low fiber content where the intent of adding fibers is merely to improve the resistance of the concrete to plastic shrinkage cracking during the first 8–12 hours. The second involves higher fiber contents where the intent is to exploit later-age improvements in properties such as post-crack residual strength and toughness (Fig. 7.43 and 7.44), impact resistance (Fig. 7.45 and 7.46) and resistance to cracking induced by restrained drying shrinkage or thermal cooling (Fig. 7.49).

In practice, and in literature and advertisements prepared by fiber manufacturers, the distinction between the two categories is not always made clear. For example, statements claiming that 0.1% by volume of certain polymeric fibers favourably affect both resistance to plastic shrinkage cracking and resistance of the hardened concrete to later-age cracking are inconsistent with the observation that in ring tests, where drying commenced at age 4 days with no possibility of cracking prior to that time because of moist curing, 0.1% of fibers had no appreciable affect on crack development compared to the control concrete without fibers (Fig. 7.49, left).

7.7.4.1. Low Fiber Content PFRC Cast-In Place Applications

These applications are primarily slab on grade where welded wire fabric (WWF) has traditionally been used to control cracking due to drying or thermal cooling. The question of whether 0.1% of polymeric fibers, usually fibrillated polypropylene or monofilament polypropylene, nylon or polyester, can effectively replace WWF as secondary reinforcement in slab-on-grade is controversial and subject to considerable debate. Many of the relevant questions have been usefully identified and discussed (Zollo and Hays, 1991). In this debate the arguments centre around what is the best choice for non-structural slab-on-grade construction, a single layer of WWF (usually at the mid-depth, but in practice often lower), distributed fiber reinforcement (usually 0.1% by volume or approximately 1 kg/m^3), or no reinforcement at all. In a comparative performance evaluation based on crack width in the ring test, 0.25% steel fiber, 0.5% polypropylene fiber, 0.5% cellulose fiber and wire mesh were judged about equal (Shah, Sarigaphuti and Karaguler, 1994).

Following the time scale from initial placement to the mature hardened state may help to establish the limits within which 0.1% of polymeric fibers can or cannot be beneficial in slab-on-grade or similar non-saturated concrete, for example, curb, sidewalk and steps (Biddle, 1991). Clearly, polymeric fibers at about 0.1% by volume help to reduce plastic shrinkage cracking under the relatively severe forced-air, low-humidity conditions typical of laboratory testing (Fig. 7.47 and 7.48). Under less severe conditions of air velocity, temperature and humidity, no cracking happens in either the control concrete without fibers or the concrete with 0.1% fibers, so this raises the question of whether the fibers are necessary, as opposed to simply desirable, for minimizing the possibility of plastic shrinkage cracking. Consequently, a claim that merits consideration is that fibers are not needed to prevent plastic shrinkage cracking when the concrete is properly proportioned and placed under the conditions of cool air and concrete temperatures, moderate humidity and low wind velocity that sometimes prevail naturally, or can usually be promoted artificially when necessary by mixture coolants, windbreaks, sunshades, fog sprays etc.. These are the conditions that constitute good concreting practice for preventing plastic shrinkage cracking. If they are met, fibers are probably unnecessary. If they are not met, fibers may well reduce or prevent plastic shrinkage cracking. It is not surprising that they have proven viable and popular in residential construction (Biddle, 1991), where substandard concreting practices are not uncommon. Cases of placement under hot, dry, windy conditions when plastic shrinkage occurred despite the presence of fibers are cited to support the view that, while fibers can help to control plastic shrinkage cracking, they are not a substitute for good concreting practice (Schupack and Stanley, 1992).

In the period following the first 12 hours after placement, either plastic shrinkage cracking has been prevented, or cracks arising from it remain for the long term posing a potential durability problem with respect to freezing and thawing or penetration of chlorides from deicing salt or seawater. The evidence from laboratory ring tests (Fig. 7.49), the relatively minute crack-bridging capability of 0.1% fibers compared with the much larger cracking stress of a typical concrete, and observations of slabs with fibers that have suffered later-age cracking in the field, support the view that 0.1% of low-modulus, creep-prone fibers cannot provide significant crack control after a crack has formed (Schupack and Stanley, 1992), a view shared by others (Portland Cement Association, 1991) (Malisch, 1986) (Guirguis and Potter, 1985). Neither can they significantly enhance load transfer by aggregate interlock across a crack which depends strongly on keeping the crack width small.

7.7.4.2. *High Fiber Content PFRC In Wet Process Shotcrete*

In the late 1980's polypropylene fibers were investigated as an alternative to steel fibers and welded wire mesh in shotcrete, and were judged viable using the wet-mix process without modifications to conventional shotcreting equipment for amounts of 4–6 kg/m^3 (0.44–0.66%) of the 38 mm fibrillated form (Morgan et al., 1989). Applications have included covering natural slate and bedrock to prevent slaking and formation of acid leachate runoff, protection of exposed mudstone from slaking during dam construction, covering waste deposits to minimize water entry and production of toxic leachates, and slope stabilization of highway rock cuts (Morgan, 1994). An incinerator waste ash deposit near Vancouver was encapsulated with a 50 mm thick layer of wet process shotcrete containing 3 kg/m^3 of fibrillated polypropylene fibers (Fig. 7.51, upper and middle) (Morgan, 1994). A 50 mm thick lining containing 6 kg/m^3 of 38 mm fibrillated polypropylene was used in part of a drainage tunnel for the Oldman Dam in Alberta (Fig. 7.51, lower) (Richardson, 1990). The shotcretes developed for some of the applications involving polypropylene fibers have contained 60–65% fly ash by weight of cementitious material, and have employed a superplasticizer to facilitate mixing and placement (Malhotra, Carette and Bilodeau, 1994) (Morgan et al., 1992). They were also air-entrained with allowance made for the reduction in air content caused by pumping and spraying.

FIGURE 7.51   Wet process fibrillated polypropylene fiber shotcrete
for encapsulation of incinerator waste ash (upper and
middle) and tunnel lining at Oldman dam (lower)
(Courtesy of Forta Corporation and AGRA Earth &
Environmental Ltd.)

Monofilament fibers identified by the manufacturer as polyolefin, which appears similar or identical to polypropylene chemically and physically, have been used extensively in wet-process shotcrete repairs to berthing facilities at the Port of Montreal (Fig. 7.52, upper and middle). Although preliminary trials indicated that the 25×0.38 mm fibers could be mixed when added in paper-wrapped bundles (Fig. 5.16), and the mixture shot at fiber contents up to 1.5% volume (13.6 kg/m^3), the amount selected for the work was 1.25% (11.4 kg/m^3) because it produced shotcrete with toughness equivalent to conventional mesh-reinforced shotcrete at the 150 mm thickness applied (Morgan, Rich and Lobo, 1998). For comparison, a steel fiber shotcrete with 0.75% (60 kg/m^3) of 38 mm deformed fiber was also included in part of the work. Both shotcretes were prepared using 400 kg/m^3 of cement and 48 kg/m^3 of silica fume as described previously in 7.6.4.6. (Morgan and Rich, 1996). They were intended to meet a specified toughness level IV with a 3.5 MPa design flexural strength at 7 days (Fig. 7.13). Both met these requirements. The polyolefin fiber shotcrete reached a strength of 5.5 MPa, while the steel fiber shotcrete reached 7.4 MPa. Increases in boiled absorption, 5.4% versus 3.7%, and permeable voids content, 13.3% versus 9.5%, may have contributed to the strength difference. The residual strength factor to 2 mm deflection was about 50% for the polyolefin fiber shotcrete, consistent with Fig. 7.44, and both shotcretes met the absolute residual strength requirements of 75% and 45% of design at deflections of 0.5 and 2.0 mm respectively (Fig. 7.52, lower) (Morgan, Rich and Lobo, 1998). After 6 months including one winter of service the polyolefin fiber shotcrete showed some fine hairline cracks up to 0.4 mm wide, mainly near points of restraint provided by anchors and conventional reinforcement, while the steel fiber shotcrete was relatively free of cracks.

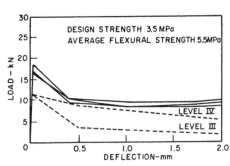

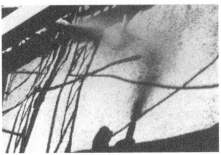

FIGURE 7.52   Deteriorated berthing facility in Port of Montreal
prior to shotcreting (upper and middle) Courtesy of
AGRA Earth and Environmental Ltd. and load-
deflection relationships for wet-process shotcrete
with 11.4 kg/m³ of polyolefin fibers (lower)
(Morgan, Rich and Lobo, 1998)

7.7.4.3. High Fiber Content Cast-In-Place and Precast PFRC

Fibrillated polypropylene fibers at 0.55% by volume (5 kg/m^3) in a 35 MPa concrete employing 415 kg/m^3 of cement and superplasticizing admixture have been used in a novel type of bridge deck design pioneered in Nova Scotia (Newhook and Mufti, 1996) that employs 200 mm thick PFRC deck slabs placed on steel girders and a connected network of welded steel strapping (Bakht and Mufti, 1996). The use of special weathering steel and PFRC steel-free deck slabs is intended to overcome traditional deck deterioration problems due to rebar corrosion. The first example of this type of construction was completed at the Salmon River crossing of the Trans Canada Highway late in 1995 (Newhook and Mufti, 1996).

In another novel application intended to exploit its improved impact resistance, 1.0% by volume of polypropylene fibers (9.1 kg/m^3) has been used in a cellular concrete to produce lightweight (560–640 kg/m^3) precast cladding panels as part of building envelopes for habitable structures exposed to wind-blown debris typical of that produced by hurricanes (Zollo and Hays, 1994) (Hays and Zollo, 1995). The 1.2 × 2.4 m panels about 200 mm thick are intended to resist penetration by debris and internal pressurization of the building by wind that explodes walls or roofs causing severe damage or total destruction in a hurricane. The panels have been evaluated for penetration resistance using a special large missile impact test in which a standard 38x89 mm wood framing unit of length 2.1 to 2.4 m is fired at the test panel at a speed of 15 m/s. They offer superior resistance to missile penetration, and are being used in new home construction in the Miami area (Hays and Zollo, 1995).

Monofilament polyolefin fibers are being evaluated in a series of highway and bridge applications constructed in 1994 and 1995 by the South Dakota Department of Transportation (Ramakrishnan and Kakodar, 1995) (Ramakrishnan and

MacDonald, 1997). A concrete bridge deck overlay with sections containing 12 and 15 kg/m^3, (1.33 and 1.66% by volume) of 50x0.63 mm fibers was placed on one half of a milled existing deck surface in August 1994 (Fig. 7.53, upper), and Jersey barrier was placed on the same bridge shortly afterwards using the same 15 kg/m^3 fiber content. The other half of the bridge overlay was placed using a conventional low-slump concrete without fibers having the same proportions as the PFRC, namely a cement content of 494 kg/m^3 and a water-cement ratio of 0.33. Test samples taken during construction indicate an average flexural strength of 5.2 Mpa with residual strength factors $R_{10,20}$ in the range 60–70 for PFRC specimens. After 24 months only minor crazing is reported with no difference between the plain and PFRC sections. On inspection after 1 week both the plain and fiber-reinforced sections of Jersey barrier exhibited numerous shrinkage cracks attributed to drying shrinkage caused by rebar restraint and the absence of contraction or construction joints over their entire length. However, cracks decreased in number and average width with increase in fiber content from 50 of average width 0.31 mm for plain concrete to 50 of average width 0.11 mm for 1.33% fibers and 38 of average width 0.06 mm for 1.66% fibers (Ramakrishnan and MacDonald, 1997).

Mixtures with the same 1.33% and 1.66% fiber content were placed in a white-topping overlay project on scarified asphalt in four slabs 100 mm thick and 16 m long without joints, but in this case the matrix contained 69 kg/m^3 of fly ash and 345 kg/m^3 of cement. Test specimens taken during construction indicate an average 28-day flexural strength of 4.8 MPa with a residual strength factor $R_{10,20}$ of 60–70. After 2 winters in service only one very narrow crack about 2 m long is reported in one of the two slabs with 1.33% fibers (Ramakrishnan and MacDonald, 1997).

A 22.9 × 4.6 m test section of full-depth pavement 140 mm thick was placed using 15 kg/m^3, 1.65% by volume, of 50 × 0.63 mm polyolefin fibers in conjunction with sections of plain

FIGURE 7.53   Bridge deck overlay placement on US 83 (upper) and full-depth highway construction (lower) in South Dakota using polyolefin fiber-reinforced concrete (Courtesy of V. Ramakrishnan)

concrete and steel fiber-reinforced concrete with 40 kg/m^3, 0.5% by volume, of 60 × 0.8 mm hooked wire fibers (Fig. 7.53, lower). The concrete matrix contained 68 kg/m^3 of fly ash with 315 kg/m^3 of cement at a water-cementitious material ratio of about 0.41. Test specimens indicate a 28-day flexural strength of 5.4 MPa with residual strength factor $R_{10,20}$ of 70–90 for the polyolefin fiber concrete. Fatigue tests to 2 million cycles show endurance limits based on static flexural strength of 67–68% for both plain and polyolefin fiber concrete, compared with 77% for the steel fiber concrete. After 9 months, including one winter in service, the sections with polyolefin fiber, steel fiber, and no fiber each have some cracks affecting only a small portion of their area that are attributed to compaction problems and subgrade conditions associated with sewer lines and solid rock outcrops (Ramakrishnan and MacDonald, 1997).

The diversity of applications investigated in South Dakota and their apparent initial success (Ramakrishnan and Kakodar, 1995) suggests that high fiber content polyolefin fiber-reinforced concrete with 1.25–1.5% fibers by volume may be viable as an alternative to steel fiber-reinforced concrete with 0.5–0.75% fibers that is typical in reconstruction or repair of pavements and overlays using fibers.

CHAPTER 8

Concluding Summary

The range of possibilities for making fiber-reinforced cements or concretes is very wide in terms of fiber composition and physical characteristics, matrix constituents, manufacturing processes, and proven or potential applications. All of the possibilities have in common the aim of using fibers to enhance the properties of an inherently weak (in tension), brittle and crack-prone cementitious matrix with respect to some or all of the following – tensile strength, flexural strength, ductility, toughness, impact and fatigue performance, resistance to cracking, and durability. Their effectiveness depends initially on achieving uniform distribution of the fibers throughout the matrix without damaging them in the process, coupled with proper consolidation of the mixture of fibers and matrix that forms the finished composite. In the short term, early-age property enhancement depends on the bond developed between fiber and matrix and the resistance to fiber pullout from the matrix. In the long term, sustained property enhancement depends on the chemical compatibility of the fiber with the matrix ingredients in terms of whether the fiber-matrix bond or fiber strength are affected by aging under various exposure conditions. Compatibility of fibers and matrix with respect to both the freshly mixed and hardened states is the key to using fibers successfully (Johnston, 1997).

The fibers available for reinforcement of cement-based matrices vary widely in composition, size, fragility, pullout resistance and consequent reinforcing effectiveness, and long-term compatibility with matrix ingredients. Most have disadvantages or limitations in one or more of these categories. The

ideal fiber yet to be developed will be characterized by high
tensile strength and elastic modulus, amenability to improvements that enhance pullout resistance without increasing
aspect ratio to the point of causing excessive mixture stiffening, ease of distribution into the manufacturing process in
amounts sufficient to significantly improve the properties of
the resulting composite, high resistance to damage by the
process, high pullout resistance approaching but not exceeding
the fiber tensile strength, and long-term chemical compatibility with the cementitious matrix and any admixtures included
in it. Often, the fiber types that best fulfil these conditions are
the most expensive, while those that are cheapest have major
disadvantages and limitations.

Bibliography

ACI Committee 211, (1988), Standard Practice for Selecting Proportions for Normal, Heavyweight and Mass Concrete, ACI Manual of Concrete Practice, MCP-1, Report 211.1, 38 pp.

ACI Committee 506, (1984), State-of-the-Art Report on Fiber-Reinforced Shotcrete, ACI Concrete International, Vol. 6, No. 12, pp. 15–27 and MCP-5.

ACI Committee 544, (1988–1), Measurement of Properties of Fiber Reinforced Concrete, MCP-5, Report 544.2R, 11 pp.

ACI Committee 544, (1988–2), Design Considerations for Steel Fiber Reinforced Concrete, MCP-5, Report 544.4R, 16 pp.

ACI Committee 544, (1993), Guide for Specifying, Proportioning, Mixing, Placing and Finishing SFRC, MCP-5, Report ACI 544.3R, 10 pp.

ACI Committee 544, (1996), Fiber-Reinforced Concrete, Report 544.1R, MCP-5, 66 pp.

ASTM, (1994), ASTM Book of Standards for Concrete and Aggregates, Vol. 04.02.

ASTM, (1995), ASTM Book of Standards for Chemical-Resistant Materials; Vitrified Clay, Concrete, Fiber-Cement Products; Mortars; Masonry, Vol. 04.05.

Adams, M., (1975), Fiber Reinforced Cement and Concrete, RILEM Symposium, Construction Press Ltd., U.K., pp. 527–531.

Aggarwal, L.K., (1992), Fibre Reinforced Cement and Concrete, RILEM Proceedings 17, E & FN Spon, pp. 1120–1127.

Aggarwal, L.K. and Singh, J., (1990), Cement and Concrete Composites, Vol. 12, No. 2, pp. 103–108.

Akers, S.A.S., Studinka, J.B., Meier, P., Dobb, M.G., Johnson, D.J. and Hisaka, J., (1989), International Journal of Cement Composites and Lightweight Concrete, Vol. 11, No. 2, pp. 79–92.

Akers, S.A.S. and Studinka, J.B., (1989), International Journal of Cement Composites and Lightweight Concrete, Vol. 11, No. 2., pp. 93–97.

Akihama, S., Suenaga, T. and Banno, T., (1982), Journal of Japan Concrete Institute, Vol. 20, No. 8, pp. 75–84.

Akihama, S., Suenaga, T. and Banno, T., (1986–1), International Journal of Cement Composites, Vol. 8, No. 1, pp. 21–33.

Akihama, S., Nakagawa, H., Takada, T. and Yamaguchi, M., (1986–2), Developments in Fibre Reinforced Cement and Concrete, RILEM Symposium Proceedings FRC 86 (Ed. Swamy et al.), Vol. 1, Paper 2.5.

Akihama, S., Suenaga, T., Tanaka, M. and Hayashi, M., (1987), Fiber Reinforced Concrete Properties and Applications, ACI Special Publication SP-105, pp. 189–209.

Akihama, S., Suenaga, T. and Nakagawa, H., (1988), ACI Concrete International Vol. 10, No. 1, pp. 40–47.

Allen, H.G., (1971), Journal of Composite Materials, Vol. 5, April, pp. 194–207.

Al-Tayyib, A.J. and Al-Zahrani, M.M., (1990), ACI Materials Journal Vol. 87, No. 2, pp. 108–113.

Ando, T., Sakai, H., Takahashi, K., Hoshjima, T., Awata, M. and Oka, S., (1990), Thin-Section Fiber Reinforced Concrete and Ferrocement, ACI Special Publication SP-124, pp. 39–60.

Andrews, G. and Sharma, A.K., (1990), ACI Concrete International, Vol. 12, No. 3, pp. 53–57.

Armelin, H.S. and Helene, P., (1995), ACI Materials Journal, Vol. 92, No. 3, pp. 258–267.

Aveston, J., Mercer, R.A., and Sillwood, J.M., (1974), National Physical Laboratory (U.K.) Conference Proceedings, pp. 93–103.

Austroads, (1992), A Guide to Structural Design of Road Pavements, Austroads, Sydney, Australia.

B.H.P. Reinforcing Products, (1994), Fibresteel Technical Manual, B.H.P. Reinforcing Products, Milperra, N.S.W. 1891, Australia.

Baggott, R. and Abdel-Monem, A.E.S., (1992), High Performance Fiber Reinforced Cement Composites, RILEM Proceedings 15, E & FN Spon, pp. 444–455.

Bagate, M., McCullough, B.F. and Fowler, D., (1985), Transportation Research Record No. 1040, Washington D.C., pp. 25–33.

Bakht, B. and Mufti, A.A., (1996), ACI Concrete International, Vol. 18, No. 2, pp. 50–55.

Balaguru, P., Narahari, R. and Patel, M., (1992), ACI Materials Journal, Vol. 89, No. 6, pp. 541–546.

Balaguru, P. and Ramakrishnan, V., (1986), ACI Journal, Proceedings Vol. 83, No. 3, pp. 374–382.

Balaguru, P., (1994), ACI Materials Journal, Vol. 91, No. 3, pp. 280–288.

Balaguru, P., (1995), Fiber Reinforced Concrete Modern Developments (Ed. Banthia and Mindess), University of British Columbia, pp. 319–334.

Balaguru, P. and Slattum, K.L., (1995), Testing of Fiber Reinforced Concrete, ACI Special Publication SP-155, pp. 115–136.

Ball, H.P. and Wackers, M., (1993), Proc. of 9th Biennial Congress, Glassfibre Reinforced Cement Association, Wigan, U.K., 16 pp.

Banthia, N., Mindess, S. and Bentur, A., (1987), Materials and Structures, RILEM, Vol. 20, No. 118, pp. 293–302.

Banthia, N. and Sheng, J., (1991), Fiber Reinforced Cementitious Materials, Materials Research Society Symposium Proceedings, Vol. 211, pp. 25–32.

Banthia, N., Trottier, J.F., Pigeon, M. and Krishnadev, M.R., (1992), High Performance Fiber Reinforced Cement Composites, RILEM Proceedings 15, E & FN Spon, pp. 456–465.

Banthia, N., Trottier, J.F., Beaupre, D. and Wood, D., (1994), ACI Concrete International, Vol. 16, No. 6, pp. 27–31.

Banthia, N., Azzabi, M. and Pigeon, M., (1993), Materials and Structures, RILEM, Vol. 26, No. 161, pp. 405–413.

Banthia, N., Moncef, A. and Sheng, J., (1994), Thin Reinforced Concrete Products and Systems, ACI Special Publication SP-146, pp. 43–68.

Banthia, N. and Trottier, J.F., (1994), ACI Materials Journal, Vol. 91, No. 5, pp. 435–446.

Banthia, N., Azzabi, M. and Pigeon, M., (1995), Testing of Fiber Reinforced Concrete, ACI Special Publication SP-155, pp. 137–152.

Banthia, N. and Trottier, J.F., (1995–1), ACI Materials Journal, Vol. 92, No. 1, pp. 48–57.

Banthia, N. and Trottier. J.F., (1995–2), ACI Materials Journal, Vol. 92, No. 2, pp. 146–154.

Banthia, N. and Dubey, A., (1996), University of British Columbia, Department of Civil Engineering, 25 pp., ACI Materials Journal, in press 2000.

Banthia, N., Mindess, S. and Trottier, J.F., (1996), ACI Materials Journal, Vol. 93, No. 5, pp. 472–479.

Baun, M.D., (1993), Maintenance of Roadway Pavement and Structures, Transportation Research Record No. 1392, Washington D.C., pp. 73–78.

Batson, G.B. and Youssef, A.G., (1994), Fibre Reinforced Concrete Developments and Innovations, ACI Special Publication SP-142, pp. 141–166.

Beckett, D., (1990), Concrete (U.K.), Vol. 24, No. 3, pp. 43–45.

Bekaert, N.V. (S.A.), (1984), Dramix^R Industrial Flooring, N.V. Bekaert S.A., Zwevegem, Belgium.

Bekaert, N.V. (S.A.), (1995), Dramix^R Steel Fiber Design Manual, N.V. Bakaert S.A., Zwevegem, Belgium.

Bekaert, N.V. (S.A.), (1996), Steel Wire Fibre Reinforced Segments for Tunnel Linings, N.V. Bekaert S.A., Zwevegem, Belgium.

Bentur, A. and Diamond, S., (1987), International Journal of Cement Composites and Lightweight Concrete, Vol. 9, No. 3, pp. 127–135.

Bentur, A. and Akers, S.A.S., (1989), International Journal of Cement Composites and Lightweight Concrete, Vol. 11, No. 2, pp. 99–116.

Bentur, A., (1990), Thin-Section Fiber Reinforced Concrete and Ferrocement, ACI Special Publication SP-124, p p. 215–231.

Bergstrom, S.G., (1975), Fiber Reinforced Cement and Concrete, RILEM Symposium, Construction Press Ltd., U.K., pp. 595–600.

Berhane, Z., (1994), Materials and Structures, RILEM, Vol. 27, No. 170, pp. 347–352.

Berke, N.S. and Dallaire, M.P., (1994), Fiber Reinforced Concrete Developments and Innovations, ACI Special Publication SP-142, pp. 19–42.

Bernard, E.S., (1997 and 1998), University of Western Sydney, Nepean, Civil Engineering Reports CE5 and CE8.

Biddle, D.T., (1991), ACI Concrete International, Vol. 31, No. 2, pp. 46–47.

Bijen, J., (1986), Durability of Glass Fiber Reinforced Concrete, Symposium Proceedings, Precast/Prestressed Concrete Inst., Chicago, pp. 251–263.

Bijen, J., (1990), International Journal of Cement Composites and Lightweight Concrete, Vol. 12, No. 2, pp. 95–101.

Bijen, J., de Haas, G. and Bosmans, G., (1994), Institute for Materials and Environmental Research B.V., Netherlands, Report 94293a, 25 pp.

Blood, G.W., (1970), M.Sc. Thesis, University of Calgary.

Broutman, L.J. and Krock, R.H., (1967), Modern Composite Materials, Addison-Wesley, 581 pp.

Building Research Establishment, (1988), Digest No. 331, GRC, Watford, U.K..

Canovas, M.F.C., Selva, N.H. and Kawiche, G.M., (1992), Materials and Structures, RILEM, Vol. 25, No. 151, pp. 417–422.

Chanvillard, G., Aitcin, P.C. and Lupien, C., (1989), International Symposium on Recent Developments in Concrete Fiber Composites, Transportation Research Record 1226, Washington D.C., pp. 48–56.

Chatveera, B. and Nimityongskul, P., (1992), Fiber Reinforced Cement and Concrete, RILEM Proceedings 17, E & FN Spon, pp. 1056–1072.

Chen, L., Mindess, S., Morgan, D.R., Shah, S.P., Johnston, C.D. and Pigeon, M., (1995), Testing of Fiber Reinforced Concrete, ACI Special Publication SP-155, pp. 41–75.

Chen, Z.Y., Cheng, Q.H. and Gao, F., (1991), Fiber-Reinforced Cementitious Materials, Materials Research Society Symposium Proceedings, Vol. 211, pp. 209–214.

Clements, M.J.K., (1996), Proceedings of IX Australian Tunneling Conference, Sydney, pp. 261–268.

Concrete Construction, (1986), Editorial, January, pp. 49.

Concrete Society, (1973), Fibre Reinforced Cement Composites, Technical Report 51.067, Concrete Society, Slough, U.K., 77 pp.

Concrete Society, (1994), Concrete Industrial Ground Floors, Technical Report No. 34, Concrete Society, Slough, U.K., 146 pp.

Cook, D.J., (1980), Fibrous Concrete, CI80, Concrete Society, Slough, U.K., Construction Press Ltd., pp. 99–109.

Coutts, R.S.P., (1989), International Journal of Cement Composites and Lightweight Concrete, Vol. 11, No. 3, pp. 143–147.

Coutts, R.S.P., (1992), Fibre Reinforced Cement and Concrete, RILEM Proceeding 17, E & FN Spon, pp. 31–47.

Craig, R.J., Parr, J.A., Germain, E., Mosquera, V. and Kamilares, S., (1986), ACI Journal, Vol. 83, No. 6, pp. 934–942.

Craig, R.J., (1987), Fiber Reinforced Concrete Properties and Applications, ACI Special Publication, SP-105, pp. 517–564.

Dahl, P.A., (1985), SINTEF Report STF65 A85039, FCB Cement and Concrete Research Institute, Trondheim, 14 pp.

Daniel, J.I. and Anderson, E.D., (1986), Developments in Fiber Reinforced Cement and Concrete, RILEM Symposium Proceedings FRC 86 (Ed. Swamy et al.), Vol. 1, Paper 2.8.

de Guillebon, B. and Sohm, J.M., (1986), Developments in Fiber Reinforced Cement and Concrete, RILEM Symposium Proceedings FRC 86 (Ed. Swamy et al.), Vol. 1, Paper 2.7.

de Lhoneux, B. and Avella, T., (1992), Fibre Reinforced Cement and Concrete, RILEM Proceedings 17, E & FN Spon, pp. 1152–1165.

Destrée, X. and Sahloul, M., (1991), Fiber Reinforced Cementitious Materials, Materials Research Society Symposium Proceedings Vol. 211, pp. 189–196.

Destrée, X., (1995), Fiber Reinforced Concrete Modern Developments (Ed. Banthia and Mindess), University of British Columbia, pp. 77–86.

Edgington, J., (1973), Steel Fibre Reinforced Concrete, Ph.D. Thesis, University of Surrey, U.K.

Edgington, J., (1977), Fibre Reinforced Materials — Design and Application, Institution of Civil Engineers Conference, London, U.K.

Edgington, J., Hannant, D.J., and Williams, R.I.T., (1974), Building Research Establishment Current Paper CP 69/74, HMSO, U.K.

EFNARC, (1996), European Specification for Sprayed Concrete, European Federation of Producers and Applicators of Specialist Products for Structures, Aldershot, U.K., 30 pp.

Esbjerg Tovvaerkstabrik A/S, (1986), Polycrete Fibre Product Brochure, Esbjerg, Denmark.

Falkner, H., Huang, Z. and Teutsch, M., (1995), ACI Concrete International, Vol. 17, No. 1, pp. 45–51.

Fanella, D.A. and Naaman, A.E., (1985), ACI Journal, Vol. 82, No. 4, pp. 475–483.

Fordos, Z. and Tram, B., (1986), Developments in Fibre Reinforced Cement and Concrete, RILEM Symposium Proceedings FRC 86 (Ed. Swamy et.al.), Vol. 1, Paper 2.9.

Forrest, M.P., Morgan, D.R., Obermeyer, J.F., Parker, P.L. and LaMoreaux, D.D., (1995), ACI Concrete International, Vol. 17, No. 11, pp. 30–36.

Fukushima, T., Shirayama, K., Hitotsuya, K. and Marayuma, T., (1992), Fibre Reinforced Cement and Concrete, RILEM Proceedings 17, E & FN Spon, pp. 965–979.

Fyles, K., Litherland, K.L. and Proctor, B.A., (1986), Developments in Fibre Reinforced Cement and Concrete, RILEM Symposium FRC 86 (Ed. Swamy, et al.), Vol. 2, Paper 7.5.

Gale, D.M., Shah, A.H. and Balaguru, P.N., (1990), Thin-Section Fiber Reinforced Concrete and Ferrocement, ACI Special Publication SP-124, pp. 61–77.

Gale, D.M., (1994), Thin Reinforced Concrete Products and Systems, ACI Special Publication SP-146, pp. 1–10.

Gilbride, P., Morgan, D.R. and Bremner, T.W., (1988), Concrete in the Marine Environment, ACI Special Publication SP-109, pp. 199–225.

Gopalaratnam, V.S. and Shah, S.P., (1986), ACI Journal, Proceedings Vol. 83, No. 1, pp. 117–126.

Gopalaratnam, V.S. and Shah, S.P., (1987), ASCE Journal of Engineering Mechanics, Vol. 113, No. 5, pp. 635–652.

Gram, H., Fagerlund, G. and Skarendahl, A., (1978), Testing and Test Methods of Fibre Cement Composites, RILEM Symposium Proceedings, Construction Press Ltd., U.K., pp. 503–509.

Gram, H.E., (1986), Developments in Fibre Reinforced Cement and Concrete, RILEM Symposium FRC 86 (Ed. Swamy et al.), Vol. 2, Paper 7.1.

Gray, R.J., (1982), Fiber-Matrix Bond Strength and its Influence on Mechanical Properties of Steel Fiber Reinforced Mortars, Ph.D. Thesis, University of Calgary, 530 pp.

Gray, R.J. and Johnston, C.D., (1987), International Journal of Cement Composites and Lightweight Concrete, Vol. 9, No. 1, pp. 43–55.

Greig, I.R.K., (1990), Thin-Section Fiber-Reinforced Concrete and Ferrocement, ACI Special Publication SP-124, pp. 233–246.

Grzybowski, M. and Shah, S.P., (1990), ACI Materials Journal, Vol. 87, No. 2, pp. 138–148.

Guirguis, S. and Potter, R.J., (1985), Technical Report TR/F90, Cement and Concrete Association of Australia, North Sydney 2060, Australia, 20 pp.

Hackman, L.E., (1980), Fibrous Concrete, CI80, Concrete Society, Slough, U.K., Construction Press Ltd., pp. 137–152.

Hähne, H., Techen, H. and Wörner, J.D. (1992), Fibre Reinforced Cement and Concrete, RILEM Proceedings 17, E & FN Spon, pp. 690–699.

Hannant, D.J., (1972), Materials and Structures, RILEM, Vol. 5, No. 25, pp. 41–44.

Hannant, D.J., (1974), Magazine of Concrete Research, Vol. 26, No. 86, pp. 47–48.

Hannant, D.J., (1978), Fibre Cements and Fibre Concretes, Wiley, 215 pp.

Hannant, D.J., (1980), Fibrous Concrete, CI80, Concrete Society, Slough, U.K., Construction Press Ltd., pp. 1–12.

Hannant, D.J. and Edgington, J., (1975), Fibre Reinforced Cement and Concrete, RILEM Symposium Proceedings, Construction Press Ltd., U.K., pp. 159–169 and 533–538.

Hanson, N.W., Roller, J.J., Daniel, J.I. and Weinmann, T.L., (1990), Thin-Section Fiber Reinforced Concrete and Ferrocement, ACI Special Publication SP-124, pp. 183–214.

Hayashi, M., Sato, S. and Fujii, H., (1986), Proc. of Symposium on Durability of Glass Fiber Reinforced Concrete, Precast/Prestressed Concrete Institute, Chicago, pp. 270–284.

Hayashi, M., Suenaga, T., Uchida, I. and Takahashi, S., (1992), Fibre Reinforced Cement and Concrete, RILEM Proceedings 17, E & FN Spon, pp. 888–901.

Hays, C.D. and Zollo, R.F., (1995), Testing Fiber Reinforced Concrete, ACI Special Publication SP-155, pp. 190–217.

Henager, C.H., (1981–1), ACI Concrete International, Vol. 3, No. 1, pp. 50–58.

Henager, C.H., (1981–2), ACI Concrete International, Vol. 3, No. 9, pp. 13–18.

Hibbert, A.P. and Hannant, D.J., (1981), Transport and Road Research Laboratory Report SR 654, Crowthorne, U.K., 25 pp.

Hikasa, J. and Genba, T., (1986), International Man-Made Fibres Congress, Austrian Chemical Institute.

Hoff, G., (1986), Steel Fiber Concrete (Ed. Shah and Skarendahl), Elsevier, pp. 67–108.

Hoff, G., (1987), Concrete Durability, ACI Special Publication SP-100, pp. 997–1041.

Homrich, J.R. and Naaman, A.E., (1987), Fiber Reinforced Concrete Properties and Applications, ACI Special Publication SP-105, pp. 283–304.

Houget, V., Ambroise, J. and Pera, J. (1995), Materials and Structures, RILEM, Vol. 28, No. 178, pp. 220–229.

Hsu, T.C., (1981), ACI Journal, Proceedings Vol. 78, No. 4, pp. 292–305.

Hubler, R.L., (1986), Engineering Digest, April, pp. 32–33.

JCI Committee, (1984), JCI Standards for Test Methods for Fiber Reinforced Concrete, Japan Concrete Institute, pp. 35–68 (in english).

JSCE Concrete Committee, (1984), Concrete Library International No. 3, Japan Society of Civil Engineers, pp. 1–74 (in english).

Jelidi, A., Chabannet, M., Ambroise, J. and Pera, J., (1991), Fiber-Reinforced Cementitious Materials, Materials Research Society Symposium Proceedings, Vol. 211 (ed. Mindess and Skalny), pp. 275–281.

Johnston, C.D., (1980), Fibrous Concrete CI80, Concrete Society, Slough, U.K., Construction Press Ltd., pp. 29–47.

Johnston, C.D., (1982), Composites, April, Butterworth, pp. 113–121.

Johnston, C.D., (1984–1), Concrete International, Vol. 6, No. 12, pp. 39–43.

Johnston, C.D., (1984–2), Cement, Concrete and Aggregates, ASTM, Vol. 6, No. 2, pp. 74–83.

Johnston, C.D., (1986), Steel Fiber Concrete, (Ed. Shah and Skarendahl), Elsevier, pp. 333–360.

Johnston, C.D., (1989), Cement, Concrete and Aggregates, ASTM, Vol. 11, No. 1, pp. 23–29.

Johnston, C.D., (1991), Fiber-Reinforced Cementitious Materials, Materials Research Symposium Proceedings, Vol. 211, (Ed. Mindess and Skalny) pp. 15–24.

Johnston, C.D., (1994–1), Significance of Tests and Properties of Concrete, ASTM Special Technical Publication STP 169C, pp. 547–561.

Johnston, C.D., (1994–2), Special Concretes Workability and Mixing, RILEM Proceedings 24 (Ed. Bartos), E & FN Spon, pp. 107–118.

Johnston, C.D., (1995–1), Testing of Fiber Reinforced Concrete, ACI Special Publication SP-155, pp. 1–22.

Johnston, C.D., (1995–2), Fiber Reinforced Concrete Modern Developments (Ed. Banthia and Mindess), University of British Columbia, pp. 87–100.

Johnston, C.D., (1996), Products Methods and Workability of Concrete, RILEM Proceedings 32 (Ed. Bartos, Marrs and Cleland), E & FN Spon, pp. 155–179.

Johnston, C.D., (1997), Advances in Concrete Technology (Ed. Malhotra), ACI Special Publication SP-171, pp. 249–278.

Johnston, C.D. and Carter, P.D., (1989), International Symposium on Recent Developments in Concrete Fiber Composites, Transportation Research Record, No. 1226, Washington, D.C., pp. 7–16.

Johnston, C.D. and Coleman, R.A., (1974), Fiber Reinforced Concrete, ACI Special Publication SP-44, pp. 177–194.

Johnston, C.D. and Skarendahl, Å., (1992), Materials and Structures, RILEM, Vol. 25, No. 148, pp. 191–200.

Johnston, C.D. and Zemp, R.W., (1991), ACI Materials Journal, Vol. 88, No. 4, pp. 374–383.

Jones, J. and Lutz, T.P., (1977), Journal of the Prestressed Concrete Institute, Vol. 22, No. 3, 25 pp.

Jorillo, P. and Shimizu, G., (1992), Fibre Reinforced Cement and Concrete, RILEM Proceedings 17, E & FN Spon, pp. 1096–1109.

Katzensteiner, B., Mindess, S., Filiatrault, A. and Banthia, N., (1994), ACI Concrete International, Vol. 16, No. 9, pp. 57–60.

Kelly, A., (1975), Fiber Reinforced Cement and Concrete, RILEM Symposium, Construction Press Ltd., U.K., pp. 463–479.

Khajuria, A., Bohra, K. and Balaguru P., (1991), Durability of Concrete, ACI Special Publication SP-126, pp. 851–868.

Kobayashi, K. and Cho, R., (1981), International Journal of Cement Composites, Vol. 3, No. 1, pp. 19–25.

Kobayashi, K., Hoshino, T. and Tsuji, K., (1990), Proceedings, Japan Society of Civil Engineers (in Japanese), Vol. 12, No. 414, pp. 195–203.

Kosa, K. and Naaman, A.E., (1990), ACI Materials Journal, Vol. 87, No. 1, pp. 27–37.

Kosa, K., Naaman, A.E. and Hansen, W., (1991), ACI Materials Journal, Vol. 88, No. 3, pp. 310–319.

Kovler, K., Sikuler, J. and Bentur, A., (1993), Materials and Structures, RILEM, Vol. 26, No. 158, pp. 231–237.

Kraai, P.O., (1985), Concrete Construction, Vol. 30, No. 9, Setember, pp. 775–778.

Krenchel, H. and Jensen, H.W., (1980), Fibrous Concrete CI80, Concrete Society, Slough, U.K., Construction Press Ltd., pp. 87–98.

Krenchel, H. and Shah, S.P., (1985), ACI Concrete International, Vol. 7, No. 3, pp. 32–34.

Krenchel, H., (1987), Nordic Concrete Research, No. 6, Oslo, pp. 143–154.

Krstulovic-Opara, N., Haghayeghi, A.R., Haidar, M. and Krauss, P., (1995), ACI Materials Journal, Vol. 92, No. 6, pp. 669–677.

Lankard, D.R., (1975), Fibre Reinforced Cement and Concrete, RILEM Symposium Proceedings, Construction Press Ltd., U.K., pp. 3–19.

Lankard, D.R., (1984), ACI Concrete International, Vol. 6, No. 12, pp. 44–47.

Lankard, D.R., (1986), Steel Fiber Concrete, (Ed. Shah and Skarendahl), Elsevier, pp. 200–217.

Lankard, D.R. and Newell, J.K., (1984), Fiber Reinforced Concrete, ACI Special Publication SP-81, pp. 287–306.

Larsen, E.T. and Krenchel, H., (1991), Fiber-Reinforced Cementitious Materials, Materials Research Society Symposium Proceedings, Vol. 211, pp. 119–124.

Lewis, G. and Mirihagalia, P., (1979), Magazine of Concrete Research, Vol. 31, No. 107, pp. 104–108.

Lewis, J., (1993), Cumberland Gap Twin Tunnel Project, Proceedings of Rapid Excavation and Tunneling Conference, Boston, U.S.A., pp. 1067–1080.

Li, V.C., Wang, Y. and Backer, S., (1990), Composites, Vol. 21, No. 2, pp. 132–140.

Linton, J.R., Berneburg, P.L., Gartner, E.M. and Bentur, A., (1991), Fiber-Reinforced Cementitious Materials, Materials Research Society Symposium Proceedings, Vol. 211, pp. 255–264.

Litherland, K.L., (1986), Proceedings of Symposium on Durability of Glass Fiber Reinforced Concrete, Precast/Prestressed Concrete Institute, Chicago, pp. 210.

Litherland, K.L., Maguire, P. and Proctor, B.A., (1984), International Journal of Cement Composites and Lightweight Concrete, Vol. 6, pp. 39.

Litherland, K.L. and Proctor, B.A., (1986), Developments in Fibre Reinforced Cement and Concrete, RILEM Symposium FRC 86 (Ed. Swamy et al.) Vol. 2, Paper 7.6.

Lyle, D.S., (1976), Modern Textiles, Wiley, New York, pp. 44.

Mai, Y.W., Andonian, R. and Cotterall, B., (1980), International Journal of Composites, Vol. 3, No. 3, pp. 149–155.

Majumdar, A.J. and Laws, V., (1991), Glass Fiber Reinforced Cement, BPS Professional Books, Division of Blackwell Scientific Publications Ltd., 197 pp.

Malhotra, V.M., Carette, G.C. and Bilodeau, A., (1994), ACI Materials Journal, Vol. 91, No. 5, pp. 478–486.

Malisch, W.R., (1986), Concrete Construction, April, Aberdeen Group, Addision, Illinois, pp. 363–368.

Malmberg, B. and Skarendahl, A., (1978), Testing and Test Methods of Fibre Cement Composites, RILEM Symposium Proceedings, Construction Press Ltd., U.K., pp. 173–179.

Mangat, P.S., (1987), Materials and Structures, RILEM, Vol. 20, No. 119, pp. 338–347.

Mangat, P.S., (1988), Materials and Structures, RILEM, Vol. 21, No. 125, pp. 352–358.

Marikunte, S. and Soroushian, P., (1994), ACI Materials Journal, Vol. 91, no. 6, pp. 607–616.

Mass, G.R., (1997), ACI Concrete International, Vol. 19, No. 6, pp. 24–27.

McGregor, G., (1995), Fiber Reinforced Concrete Modern Developments (Ed. Banthia and Mindess), University of British Columbia, pp. 63–76.

Mindess, S. and Vondran, G., (1988), Cement and Concrete Research, Pergamon, Vol. 18, pp. 109–115.

Mindess, S., Bentur, A., Yan, C. and Vondran, G., (1989), ACI Materials Journal, Vol. 86, No. 6, pp. 545–549.

Moens, J. and Nemegeer, D., (1991), ACI Concrete International, Vol. 13, No. 11, pp. 38–43.

Molloy and Associates Inc., (1994), AR Glass Fiber Tech Topics, #1, Hutchins, Texas, USA.

Molloy, H.J. and Jones, J., (1993), Proceedings of the 9th Biennial Congress of the GRCA, Glass Fibre Reinforced Cement Association, Wigan, U.K., 7 pp.

Molloy, H.J., Jones, J. and Harmon, T.G., (1993), Proceedings of the 9th Biennial Congress of the GCRA, Glass Fiber Reinforced Cement Association, Wigan, U.K..

Molloy, H.J., Jones, J. and Harmon, T.G., (1994), Thin Reinforced Concrete Products and Systems, ACI Special Publication SP-146, pp. 76–90.

Mondragon, R., (1984), SIFCON Bridge Repairs, NMERI-42, New Mexico Engineering Research Institute, Albuquerque, New Mexico.

Morgan, D.R., (1988), ACI Concrete International, Vol. 10, No. 1, pp. 24–32.

Morgan, D.R., (1991), ACI Concrete International, Vol. 13, No. 11, pp. 56–64.

Morgan, D.R., (1994), Advances in Concrete Technology (Ed. Malhotra), CANMET, Natural Resources Canada, pp. 675–720.

Morgan, D.R., (1995), Technical Reports 3, 4 and 5 to 3 M Construction Markets Division, AGRA Earth and Environmental, Burnaby, British Columbia.

Morgan, D.R., McAskill, N., Richardson, B.W. and Zellers, R.C., (1989), International Symposium on Recent Developments in Concrete Fiber Composites, Transportation Research Record No. 1226, Washington D.C., pp. 78–87.

Morgan, D.R., and McAskill, N., (1984), ACI Concrete International, Vol. 6, No. 12, pp. 33–38.

Morgan, D.R., McAskill, N., Carette, G.C. and Malhotra, V.M., (1992), ACI Materials Journal, Vol. 89, No. 2, pp. 169–177.

Morgan, D.R., Chen, L. and Beaupré, D., (1995), Engineering Foundation Conference Proceedings, Shotcrete for Underground Support, Buchen-Telfs, Austria.

Morgan, D.R., Mindess, S. and Chen, L., (1995), Fiber Reinforced Concrete Modern Developments (Ed. Banthia and Mindess) University of British Columbia, pp. 29–50.

Morgan, D.R. and Rich, L.D., (1996), ACI/SCA International Conference on Sprayed Concrete, Edinburgh, Scotland.

Morgan, D.R., Rich, L. and Lobo, A. (1998), ACI Concrete International, Vol. 20, No. 9, pp. 66–73.

NSW Roads and Traffic Authority, (1996), Concrete Roundabout Pavements – A Guide to their Design and Construction, New South Wales, Australia, 28 pp.

Naaman, A.E., (1985), ACI Concrete International, Vol. 7, No. 3, 21–25.

Naaman, A.E., (1992), High Performance Fiber Reinforced Cement Composites, RILEM Proceedings 15, E & FN Spon, pp. 18–38.

Naaman, A.E., Otter, D. and Najm, H., (1991), ACI Materials Journal, Vol. 88, No. 6, pp. 603–612.

Naaman, A.E. and Najm, H. (1991), ACI Materials Journal, Vol. 88, No. 2, pp. 135–145.

Naaman, A.E. and Shah, S.P., (1975), Fiber Reinforced Cement and Concrete, RILEM Symposium, Construction Press Ltd., U.K., pp. 171–178.

Naaman, A.E. and Shah, S.P., (1976), ASCE Proceedings, Structural Division, Vol. 102, No. ST-8, pp. 1537–1547.

Nagabhushanam, M., Ramakrishnan, V. and Vondran, G., (1989), International Symposium on Recent Developments in Concrete Fiber Composites, Transportation Research Record 1226, Washington D.C., pp. 36–47.

Nanni, A., (1990), ACI Materials Journal, Vol. 87, No. 6, 556–564.

Nanni, A. and Meamarian, N., (1991), Cement and Concrete Composites, Vol. 13, No. 2, pp. 107–114.

Narayan, R. and Kareem-Palanjian, A.S., (1983), International Journal of Cement Composites and Lightweight Concrete, Vol. 5, No. 4, pp. 235–246.

Nemegeer, D.E. and Tatnall, P.C., (1995), Testing Fiber Reinforced Concrete, ACI Special Publication SP-155, pp. 77–92.

Newhook, J.P. and Mufti, A.A., (1996), ACI Concrete International, Vol. 18, No. 6, pp. 30–34.

Nielson, N.H., (1994), Special Concretes Workability and Mixing (Ed. Bartos), RILEM Proceedings 24, E & FN Spon, pp. 69–72.

Nishioka, K., Yamakawa, S. and Shirakawa, K., (1986), Developments in Fibre Reinforced Cement and Concrete, RILEM Symposium FRC 86 (Ed. Swamy et al.), Vol. 1, Paper 2.2.

Norwegian Concrete Association, (1993), Guidelines for Shotcrete, Publication No. 7.

Oakley, D.R. and Unsworth, M.A., (1978), Testing and Test Methods of Fiber Cement Composites, RILEM Symposium, Construction Press Ltd., U.K., pp. 233–241.

Odler, I., (1988), Bonding in Cementitious Composites (Ed. Mindess and Shah), Materials Research Society Symposium Proceedings, Vol. 114, pp. 153–158.

Oesterle, R.G., Schultz, D.M. and Glikin, J.D., (1990), Thin-Section Fiber Reinforced Concrete and Ferrocement, ACI Special Publication SP-124, pp. 157–182.

Ohama, Y., Amano, M. and Endo, M., (1985), ACI Concrete International, Vol. 7, No. 3, pp. 58–62.

Oliveira, M.J.E. and Agopyan, V., (1992), Fibre Reinforced Cement and Concrete, RILEM Proceedings 17, E & FN Spon, pp. 1073–1079.

Opsahl, O.A., Buhre, K. and Hornfeldt, R., (1986), Steel Fiber Concrete (Ed. Shah and Skarendahl), Elsevier, pp. 51–66.

PCI Committee, (1991), Manual for Quality Control for Plants and Production of Glass Fiber Reinforced Concrete Products, MNL-130, Precast/Prestressed Concrete Institute, Chicago, 168 pp.

PCI Committee, (1993), Recommended Practice for Glass Fiber Reinforced Concrete Panels, MNL-128, Precast/ Prestressed Concrete Institute, Chicago, 99 pp.

Packard, R.G. and Ray, G.K., (1984), Fiber Reinforced Concrete, ACI Special Publication SP-81, pp. 325–350.

Padron, I. and Zollo, R.F., (1990), ACI Materials Journal, Vol. 87, No. 4, pp. 327–332.

Park, S.B. and Lee, B.I., (1991), Fiber-Reinforced Cementitious Materials, Materials Research Society Symposium Proceedings, Vol. 211, pp. 247–254.

Parker, F. Jr., (1974), Steel Fibrous Concrete for Airport Pavement Applications, U.S. Federal Aviation Administration, Report FAA-RD-74-31, or National Technical Information Centre, Springfield, VA, Report ADA-003-123.

Parker, F. and Rice, J., (1977), Proceedings, International Conference on Concrete Pavement Design, Purdue University, pp. 541–557.

Paul, S.L., (1976), U.S. Dept. of Commerce, NTIS Report PB-26731.

Pech, R. and Schacher, B., (1992), Fibre Reinforced Cement and Concrete, RILEM Proceedings No. 17, E & FN Spon, pp. 713–722.

Peiffer, G. and Soukatchoff, P., (1994), Special Concretes – Workability and Mixing, RILEM Proceedings 24, E & FN Spon, pp. 89–97.

Peter, I.D., (1994), Special Concretes-Workability and Mixing, RILEM Proceedings 24, E & FN Spon, pp. 73–79.

Petersson, O., (1992), Fiber Reinforced Cement and Concrete, RILEM Proceedings No. 17, E & FN Spon, pp. 723–732.

Portland Cement Association, (1991), Fiber Reinforced Concrete, Report SP 039.01 T (Ed. Daniel), Portland Cement Association, Skokie, Illinois, 48 pp.

Proctor, B.A., (1986), Proceedings of Symposium on Durability of Glass Fiber Reinforced Concrete, Precast/ Prestressed Concrete Institute, Chicago, pp. 64.

RILEM Committee 19-FRC, (1977), Fibre Concrete Materials, Materials and Structures, RILEM, Vol. 10, No. 56, 103–120.

RILEM Committee 49-TFR, (1984), Testing Methods for Fibre Reinforced Cement-Based Composites, Materials and Structures, RILEM, Vol. 17, No. 102, pp. 441–456.

Rahimi, M. and Kesler, C.E., (1979), Journal of Structural Division, ASCE, Vol. 105, No. ST1, pp. 101–109.

Rahimi, M. and Cao, H.T., (1990), Thin-Section Fiber Reinforced Concrete and Ferrocement, ACI Special Publication SP-124, pp. 265–278.

Ramakrishnan, V., (1995), Fiber Reinforced Concrete Modern Developments (Ed. Banthia and Mindess), University of British Columbia, pp. 101–115.

Ramakrishnan, V. and Kakodkar, S., (1995), South Dakota Department of Transportation, Interim Report SD 94-04, 300 pp.

Ramakrishnan, V. and MacDonald, C.N., (1997), Durability of Concrete, ACI Special Publication SP-170, Vol. I, pp. 665–680.

Ramakrishnan, V., Speakman, J., Kakodar, S. and Sure, V.R., (1994), Transportation Research Record, No. 1458, Washington D.C., pp. 48–56.

Ramakrishnan, V., Wu, G.Y. and Hosalli, G., (1989), International Symposium on Recent Developments in Concrete Fiber Composites, Transportation Research Record No. 1226, Washington D.C., pp. 17–24.

Rettberg, W.A., (1986), Hydro-Review, Spring, pp. 18–22.

Richardson, B.W., (1990), Concrete Construction, January, pp. 33–35.

Robins, P.J. and Austin, S.A., (1986), Developments in Fibre Reinforced Cement and Concrete, RILEM Symposium FRC 86 (Ed. Swamy et al.), Vol. 1, Paper 3.7.

Robinson, C., Colasanti, A. and Boyd, G., (1991), ACI Concrete International, Vol. 13, No. 4, pp. 30–35.

Rotondo, P.L. and Weiner, K.H., (1986), ACI Concrete International, Vol. 8, No. 12, pp. 22–27.

Rossi, P. and Chanvillard, G., (1992), High Performance Fiber Reinforced Cement Composites, RILEM Proceedings 15, E & FN Spon, pp. 129–139.

Ryder, J.F., (1975), Fiber Reinforced Cement and Concrete, RILEM Symposium, Construction Press Ltd., U.K., 23–35.

Sakai, K., Ochi, T. and Kitoh, M., (1995), Fiber Reinforced Concrete Modern Developments (Ed. Banthia and Mindess), University of British Columbia, pp. 51–62.

Sakai, H., Takahashi, K., Mitsui, Y., Ando, T., Awata, M., and Hoshijima, T., (1994), Fiber Reinforced Concrete Developments and Innovations, ACI Special Publication SP-142, pp. 121–140.

Sandell, N. and Westerdahl, B., (1986), Steel Fiber Concrete (Ed. Shah and Skarendahl), Elsevier, pp. 25–40.

Sanjuan, M.A., Moragues, B., Baele, B. and Andrade, C., (1991), Fiber Reinforced Cementititous Materials,

Materials Research Society Symposium Proceedings, Vol. 211, pp. 71–77.

Savastano, H. and Agopyan, V., (1992), Fibre Reinforced Cement and Concrete, RILEM Proceedings 17, E & FN Spon, pp. 1110–1119.

Schneider, B., (1992), High Performance Fiber Reinforced Cement composites, RILEM Proceedings 15, E & FN Spon, pp. 177–194.

Schraeder, E., (1986), Steel Fiber Concrete (Ed. Shah and Skarendahl), Elsevier, pp. 110–131.

Schraeder, E.K., (1988), International Committee on Large Dams, Bulletin No. 40.

Schupack, M., (1986), Steel Fiber Concrete (Ed. Shah and Skarendahl), Elsevier, pp. 479–496.

Schupack, M. and Stanley, W.R., (1992), ACI Concrete International, Vol. 14, No. 2, pp. 50–56.

Schürhoff, H.J. and Gerritse, A., (1986), Developments in Fibre Reinforced Cement and Concrete, RILEM Proceedings FRC 86 (Ed. Swamy et al.), Vol. 1, Paper 2.6.

Sethunarayan, R., Chockalingham, S. and Ramanathan, R., (1989), International Symposium on Recent Developments in Concrete Fiber Composites, Transportation Research Record, No. 1226, Washington D.C., pp. 57–60.

Shaeles, E.A. and Hover, K.C., (1988), ACI Materials Journal, Vol. 85,No. 6, pp. 495–504.

Shah, S.P., Karaguler, M.E., and Sarigaphuti, M., (1992), ACI Materials Journal, Vol. 89, No. 3, pp. 289–295.

Shah, S.P., Ludirija, D., Daniel, J.I. and Mobasher, B., (1988), ACI Materials Journal, Vol. 85, No. 5, pp. 352–360.

Shah, S.P., Sarigaphuti, M. and Karaguler, M.E., (1994), Fiber Reinforced Concrete Developments and Innovations, ACI Special Publication SP-142, pp. 1–18.

Shah, S.P., Stroeven, P., Dalhuisen, D. and van Steckelenberg, P., (1978), Testing and Test Methods for Fiber Cement Composites, RILEM Symposium Proceedings, Construction Press Ltd., U.K., pp. 399–408.

Shao, Y., Marikunte, S. and Shah, S.P., (1995), ACI Concrete International, Vol. 17, no. 4, pp. 48–53.

Shao, Y. and Shah, S.P., (1997), ACI Materials Journal, Vol. 94, No. 6, pp. 555–564.

Simatupang, M.H. and Lange, H., (1987), International Journal of Cement Composites and Lightweight Concrete, Vol. 9, No. 2, pp. 109–112.

Skarendahl, Å., (1992), High Performance Fiber Reinforced Cement Composites, RILEM Proceedings 15, E & FN Spon, pp. 156–163.

Skarendahl, Å., (1994), Proceedings, International Symposium on Brittle Matrix Composites (Ed. Brandt et al.), IKE and Woodhead Publishing, Warsaw, pp. 44–50.

Skatun, O. and Spigerverk, C., (1986), Developments in Fibre Reinforced Cement and Concrete, RILEM Symposium FRC 86, (Ed. Swamy et al.), Vol. 2, Paper 9.6.

Smith, R.T., (1991), Concrete (U.K.) Vol. 25, No. 2, pp. 14–15.

Smith, R. and Wilde, R.E., (1996), ACI Concrete International, Vol. 18, No. 6, pp. 35–38.

Sobral, H.S., (Editor)(1990), Vegetable Plants and their Fibres as Building Materials, RILEM Proceedings 7, E & FN Spon, 392 pp.

Soroushian, P., Aouadi, F. and Nagi, M., (1991), ACI Materials Journal Vol. 88, No. 1, pp. 11–18.

Soroushian, P., Bayasi, Z. and Khan, A., (1990), Thin-Section Fiber Reinforced Concrete and Ferrocement, ACI Special Publication SP-124, pp. 79–98.

Soroushian, P. and Marikunte, S., (1990), Thin-Section Fiber Reinforced Concrete and Ferrocement, ACI Special Publication SP-124, pp. 99–124.

Soroushian, P., Khan, A. and Hsu, J., (1992), ACI Materials Journal, Vol. 89, No. 6, pp. 535–540.

Soroushian, P., Marikunte, S. and Won, J.P., (1995), ACI Materials Journal, Vol. 92, No. 2, pp. 172–180.

Soroushian, P., Mirza, F. and Alhozaimy, A., (1995), ACI Materials Journals, Vol. 92, No. 5, pp. 553–560.

Soroushian, P., Nagi, M. and Hsu, J., (1992), ACI Materials Journal, Vol. 89, No. 3, pp. 267–276.

Soroushian, P., Shah, Z. and Marikunte, S., (1994), Thin Reinforced Concrete Products and Systems, ACI Special Publication SP-146, pp. 25–42.

Soroushian, P., Tlili, A., Yohena, M. and Tilsen, B.L., (1993), ACI Materials Journal, Vol. 90, No. 1, pp. 40–49.

Studinka, J.B., (1989), International Journal of Cement Composites and Lightweight Concrete, Vol., No. 2, pp. 73–78.

Swamy, R.N., (Editor)(1988), Natural Fibre-Reinforced Cement and Concrete, Concrete Technology and Design, Vol. 5, Blackie, Glasgow.

Swamy, R.N. and Stavrides, H., (1975), Fibre Reinforced Cement and Concrete, RILEM Symposium, Construction Press Ltd., U.K., pp. 197–208.

Swamy, R.N. and Stavrides, H., (1979), ACI Journal, Proceedings Vol. 76, No. 3, pp. 443–460.

Tatnall, P.C. and Kuitenbrouwer, L., (1992), ACI Concrete International, Vol. 14, No. 12, pp. 43–47.

Tatro, S.B., (1985), Master's Thesis, Purdue University, Lafayette, Indiana, U.S.A.

Tatro, S.B., (1987), Concrete and Concrete Construction, Transportation Research Record No. 1110, Washington D.C., pp. 129–137.

Tegos, I.A., (1989), ACI Structural Journal, Vol. 86, No. 4, pp. 473–482.

Thiery, J., Vautrin, A. and Francois-Brazier, J., (1991), Fiber Reinforced Cementitious Materials, Materials Research Society Symposium Proceedings, Vol. 211, pp. 79–91.

United States Patent Office, (1969), No. 3,429,094 and 3,500,728 (1970), to Battelle Memorial Institute and 3,650,785 (1972), to U.S. Steel Corporation.

van der Plas, C., Yue, B. and Bijen, J., (1992), Fly Ash, Silica Fume, Slag and Natural Pozzolans in Concrete, Proceedings of Fourth CANMET/ACI International Conference, Supplementary Papers Natural Resources Canada, Ottawa, Ontario, pp. 175–188.

van Mier, J.G.M. and Timmers, G., (1992), Fiber Reinforced Cement and Concrete, RILEM Proceedings 17, EF&N Spon, pp. 245–256.

Vandenberghe, M.P. and Nemegeer, D.E., (1985), ACI Concrete International, Vol. 7, No. 3, pp. 54–57.

Vandenberghe, M.P., (1992), High Performance Fiber Reinforced Cement Composites, RILEM Proceedings 15, E & FN Spon, pp. 226–234.

Vandewalle, M., (1990), Tunneling the World, N.V. Bakaert S.A., Zwevegem, Belgium, 229 pp.

Vandewalle, L., (1997), Fourth CANMET/ACI International Conference on Durability of Concrete, Supplementary Papers Volume, Natural Resources Canada, Ottawa, Ontario, pp. 77–94.

Vinson, K.D. and Daniel, J.I., (1990), Thin-Section Fiber Reinforced Concrete and Ferrocement, ACI Special Publication SP-124, pp. 1–18.

Walton, P.L. and Majumdar, A.J., (1978), Building Research Establishment, Watford, U.K., Report CP 57/78.

Wang, Y., Backer, S. and Li, V.C., (1987), Journal of Materials Science, Vol. 22, pp. 4281–4291.

Wang, Y., Li, V.C. and Backer, S., (1990), ACI Materials Journal, Vol. 87, No. 5, pp. 461–468.

Wörner, J.D. and Techen, H., (1994), Special Concretes Workability and Mixing (Ed. Bartos), RILEM Proceedings 24, E & FN Spon, pp. 81–87.

Wu, G.Y. and Jones, M.P., (1987), Fiber Reinforced Concrete Properties and Applications, ACI Special Publication SP-105, pp. 403–418.

Yoda, K. and Suenaga, T., Tamaki, T. and Morimoto, J., (1992), Fibre Reinforced Cement and Concrete (Ed. Swamy), RILEM Proceedings 17, E & FN Spon, pp. 980–987.

Zellers, R.C. and Ramakrishnan, V., (1994), Concrete Research, Transportation Research Record No. 1458, Washington D.C., pp. 57–66.

Zhang, M.H., Chevrier, R.L., Malhotra, V.M. and Mirza, J., (1997), Durability of Concrete, ACI Special Publication SP-170, Vol. I, pp. 681–722.

Zollo, R.F. and Hays, C.D., (1991), ACI Concrete International, Vol. 13, No. 11, pp. 50–55.

Zollo, R.F., Ilter, J.A. and Bouchacourt, G.B., (1986), Developments in Fibre Reinforced Cement and Concrete, RILEM Symposium FRC 86 (Ed. Swamy et al.), Vol. 1, Paper 4.5.

Zollo, R.F. and Hays, C.D., (1994), ACI Concrete International, Vol. 16, No. 6, pp. 23–26.

Zonsveld, J.J., (1975), Fibre Reinforced Cement and Concrete, RILEM Symposium Proceedings, Construction Press Ltd., U.K., pp. 217–226.

INDEX

A

Printed and bound by CPI Group (UK) Ltd, Croydon, CR0 4YY

01/11/2024

01782621-0001